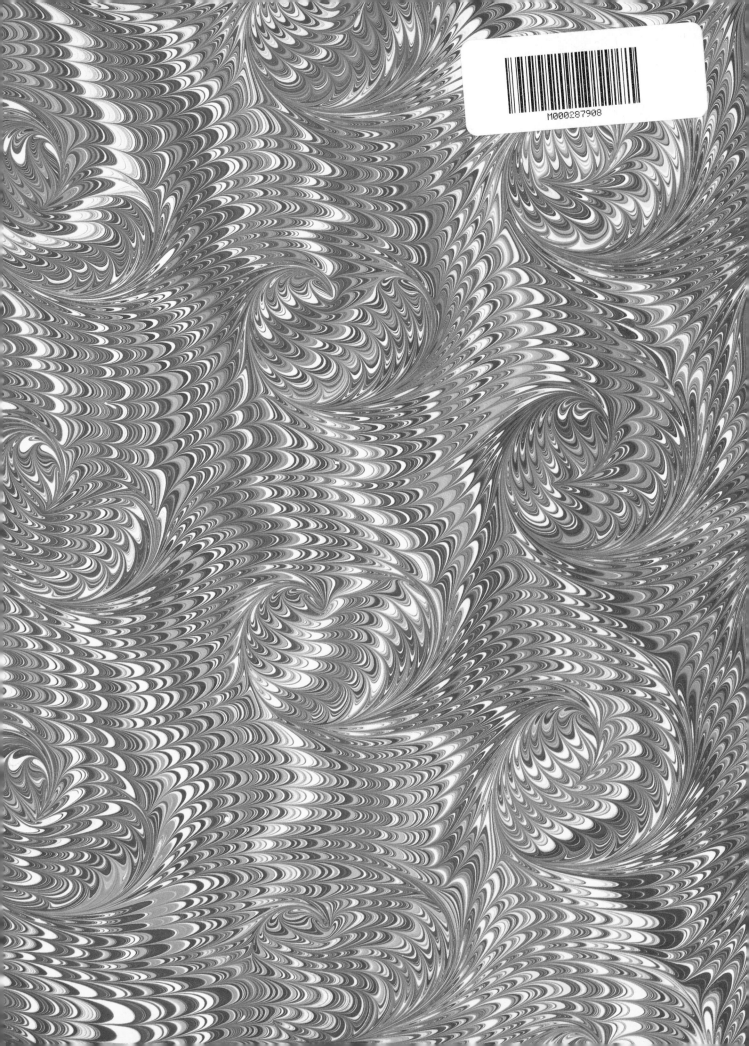

# JAPANESE TOYS

## AMUSING PLAYTHINGS FROM THE PAST

Schiffer Publishing Ltd

4880 Lower Valley Road, Atglen, PA 19310 USA

**William C. Gallagher**

Library of Congress Cataloging-in-Publication Data

Gallagher, William C.
      Japanese toys: amusing playthings from the past.
            p. cm.
Includes bibliographical references (p. ).
      ISBN 0-7643-1129-8
1. Toplay--Catalogs. 2. Tin toys--Collectors and collecting--
Japan--Catalogs. I. Title.
      NK8454.2.J34 T662 2000
      688.7'2'0952075--dc21

                                                    00-008750

Book Design by Anne Davidsen
Type set in Boink/ Humanst 521

ISBN: 0-7643-1129-8

Printed in China
1 2 3 4

Published by Schiffer Publishing Ltd.
4880 Lower Valley Road
Atglen, PA 19310
Phone: (610) 593-1777; Fax: (610) 593-2002
E-mail: Schifferbk@aol.com
Please visit our web site catalog at
**www.schifferbooks.com**

We are always looking for people to write books on new
and related subjects. If you have an idea for a book, please
contact us at the above address.

This book may be purchased from the publisher.
Include $3.95 for shipping.
Please try your bookstore first.
You may write for a free catalog.

In Europe, Schiffer books are distributed by:
Bushwood Books
6 Marksbury Ave.
Kew Gardens
Surrey TW9 4JF
England
Phone: 44 (0)208 392-8585
Fax: 44 (0)208 392-9876
E-mail: Bushwd@aol.com
Free postage in the UK. Europe: air mail at cost.

# CONTENTS

# DEDICATION

To Three G Hobbies, the family toy business which evolved from trains to toys and became the foundation of my fascination with toys and the tin windup toys from T.P.S., and to the child in all of us, no matter what your age is or what you enjoy or collect.

# ACKNOWLEDGMENTS

Even though I have been in the antique and collectible business for close to twenty years, I did not start seriously collecting T.P.S. toys until 1994. As I began to try to learn more, the first real reference came from **Don Hultzman** in Richard O'Brien's *Collecting Toys*. **Herb Smith** has also provided separate listings of T.P.S. toys in his *Smith House Toys Price Guides*. I have scoured many reference books both in the US and Japan in an effort to identify and document the toys produced. I am grateful to **Ron Smith** who specializes in Japanese tin vehicles, for his advice and insight.

It was meeting **Yoshio Udagawa,** Chairman and one of the original founders of Toplay Ltd., that convinced me to share what I have learned through this book. Without his creativity and efforts, there would be no T.P.S. toys to write about nor any catalogs, drawings, or recollections to pass on to the collector. **Kazuhiko Udagawa,** President of Toplay Ltd. and the son of the founder, was also very helpful in providing information and English translation during my research efforts.

Thanks to my many Japanese friends who have helped me find my way around Japan, with language and communications, with museum historical records, and with Japanese to English translations: **Kozo Wakabayashi, Hideki Fujimoto, Masamichi Inoue, Michimasa Yamamoto, Ike Ogawa, Kousaku Yoshioka, Ayako Honda, Miki and Atsushi Hasegawa, Kaz Tomozoe, Shigeru Mozuka** of Japan Toys Museum Foundation, and **Toyoji Takayama** of the Kyoto Tin Toy Museum.

And thanks to those who have contributed toys and photographs: **Tim Hannum, Don Hultzman, Randy Ibey** of Randy's Toy Shop, **Carl Johnson and Karen Delp, Debby and Marty Krim** of New England Auction Gallery, **John Mendoza, Dick Rowe, Brynne and Scott Shaw, Herb Smith** of Smith House Toys, and **Pete Thompson.**

Finally, thanks to my editors, **Doug Congdon-Martin** and **Donna Baker** for their help and guidance during this book building process.

# FOREWORD

I refer to these toys as "happy toys" because as you observe the wonderful designs of Yoshio Udagawa, you are taken by their whimsical and comical subjects combined with great actions, plus their overall ability to bring a smile to your face. His toys were typically not toys of war or toys of violence, but rather toys that would bring to mind happy and peaceful times. Mr. Udagawa did this intentionally. In the 1950s, the world was in a period of reconstruction and optimism. Why not have happy toys? Plus, you can see this same attitude in his personality. Mr. Udagawa is a happy person with a big smile, which he exhibited often as we reviewed toys he had forgotten or had not seen for a long time, and a person who genuinely loves and enjoys toys. I had the opportunity to set up and sell at an antique toy show in Tokyo and invited Mr. Udagawa to attend with me. It was such a pleasure to walk the aisles with him and watch his face as he recognized his toy designs from the past.

As I began documenting the toys that originated within Toplay Ltd., I became aware of how often the word "happy" was included in the name of the toy. The circus was a favorite subject of Mr. Udagawa. He created circus parades, animals, and clowns, all designed with the happy thoughts associated with this wonderful family event. Toys related to amusement parks were also happy subjects. Over fifty of the company's toys are associated with these subjects.

I was a child when the toys from Japan began to populate the 5 & 10¢ stores in the United States and what I remember is that they were funny and made you smile. I do not recall having a windup toy from T.P.S. as a child. So it is ironic to me that as an adult, almost forty years later, I have been busy collecting and researching T.P.S. toys meant for children.

Seals practicing for the opening act.

There is very strong interest in T.P.S. toys, but no one place to go for information about the company and its products. I hope this book will provide that resource and that you will have as much fun as I did learning more about the toys from Toplay (T.P.S.) Ltd.

Mr. Udagawa is nearly seventy-five years old as of this writing, and through a translator I found that there is still much of the company history, spirit, and creativity in his head. He is excited about the interest in his toys and the notion that a publication would document his company and the toys it produced.

Everyone loves the circus!

# MESSAGE FROM YOSHIO UDAGAWA

*Founder and Chairman, Toplay (T.P.S.) Ltd.*
(as translated from Japanese)

I met Mr. Gallagher for the first time in the summer of 1998 in a Tokyo hotel. I found him to be a very serious collector of Toplay tin toys and was very much impressed and surprised that he had the idea to publish a book about Toplay tin toys.

I knew Japanese collectors who had been collecting Toplay tin toys, but it was unexpected that collectors of T.P.S. were also found in the US.

I met Mr. Gallagher the second time in January of 1999. I knew that he was steadily making preparations to complete the book and how seriously he was proceeding with it. I saw his desire and the vision in his eye and asked him a few questions:

1. Why did you choose Toplay among so many Japanese makers?
2. Since the toy collection is not your main job and this is an expensive hobby, doesn't your wife complain about your hobby as a collector?

He answered those questions as follows:

1. I see a dream in Toplay tin toys that is very peaceful and unique and having an orientation which does not exist in other maker's toys. I like them and respect them for that.
2. My wife understands, supports, and shares my hobby. While I am staying in Japan, sometimes my wife is bidding on Toplay toys at auction and having cross contact with me.

When I first met him, I was surprised to note that he knew the history of Toplay tin toys much more than I did. I am ashamed that the makers easily forget the tin toys of the past. So when he showed me pictures of Toplay tin toys, I did not remember 30% of them.

You may not believe it, but when I returned to my company office after meeting with him, I talked to employees who had been with Toplay a long time. Even the employees could not remember some of the toys. When I went home that day, memories about some of them came back, and

gradually since then I have remembered more. Honestly speaking, it was really shocking but also very pleasing remembering the past.

Another thing of which I am ashamed is we did not keep samples (even one piece) at Toplay. Experiencing the mass production period may have made us forget the importance of one piece.

When this book is published, I believe it will become a very precious thing for Toplay. But also it will become a good treasure for the future. I am sure it will be a wonderful thing for Mr. Gallagher, myself, and anyone concerned with Japanese tin toys. Lastly, I would like to express my heartfelt congratulations and appreciation for this book.

Thank you very much,
Yoshio Udagawa, Chairman Toplay (T.P.S.) Ltd.

Yoshio Udagawa

# INTRODUCTION

Toys from Japan flooded the market during the post war years. They were marked "Made in Japan," but often unnoticed was the manufacturer or other marks on the toy or box. Later, as people started to collect toys and relive their childhood, there was more interest in identifying the toy by its mark. However, some toys contained multiple marks and some contained no mark, creating a lot of confusion as to the manufacturer of the toy. Trademarks will be covered later in this book, but over the years, many collectors have seen the T.P.S. mark of Toplay Ltd. on toys and have associated this mark with unique, quality, and very desirable toys. Although there were hundreds of Japanese toy manufacturers whose marks have been documented, very little history has been written about specific companies and about the interconnections that existed between companies within the toy industry in Japan.

Because of the recognition of Toplay as one of the best Japanese tin toy manufacturers from the 1950s, 1960s and 1970s, this book is focused on Toplay (T.P.S.) Ltd., owner of the T.P.S. mark and supplier to many other well known toy manufacturers and importers. The initials T.P.S. were de-rived from the formal Japanese name of Tokyo Plaything Shokai, as was the shortened company name of Toplay, pronounced toe (from the name Tokyo) -play.

When Yoshio Udagawa was asked about the origin of the three-fingered symbol that incorporates T.P.S. into a trademark, he responded by holding up three fingers and indicated that there were three people who started the company together. The trademark was then symbolic of both the company name and the three co-founders of the company.

T.P.S. trademark.

## HISTORICAL OVERVIEW

In 1954, thirty-year-old Yoshio Udagawa was working for the Kokyu Shokai toy company. He enjoyed the toy business and had many design and subject ideas, so he decided to start his own company along with an engineer, the late Yukio Miyakawa, and a designer, Yasuo Kushida. These three gentlemen were the core of the company and over the years, each of these individuals contributed his own special skills to create some of the most interesting and unique toy designs in the industry at that time. Their complex, comical, and somewhat whimsical designs captured the imagination and attention of toy buyers then and continue to attract collectors today.

Kokyu trademark.

The first toy produced by T.P.S. was the hobo "Clown on Roller Skates." This toy was an instant success and a new toy company was born. Every toy company hopes for a hit with a new toy, and fortunately the "Clown on Roller Skates" turned out to be one of Toplay's most successful toys, eventually selling over a million pieces. Even with a production quantity which was very high for the time, the realistic action of this colorful toy still makes it highly sought after all over the world. The skating theme was so popular that it became the foundation for ten different variations of the original toy.

This success was followed by the well known "Gay 90's Cyclist" and then "Skippy the Tricky Cyclist." These first three outstanding toys set the pace for a company that is still in business today!

Toy designs begin with ideas. As Mr. Udagawa and his two partners generated ideas, he or Mr. Kushida would create the design and Mr. Miyakawa would engineer the action mechanisms and turn the designs into manufacturable toys. Mr. Miyakawa has since passed away and Mr. Kushida went on to become the President of Mikuni, another well known Japanese toy company.

Left: First T.P.S. toy. Right: Re-creation of first toy.

Re-creation signed by Y. Kushida.

Note: After his retirement from Mikuni, Mr. Kushida hand made several large, 7.5" size re-creations of toys, including the roller skating Clown, roller skating Popeye, and roller skating Mickey Mouse. Approximately ten pieces of each were made. These handmade toys have been sometimes incorrectly described as prototype samples. Instead they were made more than thirty years after the original toys to commemorate their popularity. Most of these re-creations are still in Japan in the hands of major toy collectors. They can be seen at both the Takayama Museum in Kyoto and the Kitahara Museum in Yokohama.

One of the ongoing challenges facing Toplay and other Japanese toy makers was the target selling price for the toy and what that meant for the target manufacturing cost. Importers wanted toys that retailed for less than $1.00, as evidenced by the number of toy boxes marked with 79¢ and 89¢ price tags. In fact, I have a "Skating Chef" with a price tag of 39¢. Relative to today's values, it is hard to imagine manufacturing a roller skating toy to wholesale for 20¢. Toplay continually tried to add to the complexity and interest of the toy, which resulted in higher manufacturing costs. When this additional complexity, such as in the "Champ on Ice" bear skating trio, drove the final selling price above $1.00, sales for that toy declined. However, as we look back at these toys today, it is that extra action or complexity which increases the value and the desirability for current collectors.

# 1950S

Japanese mechanical windup toys were at their peak of popularity during the late 1950s and early 1960s. It was during this period that T.P.S. produced their classic and most well-recognized toys. In 1956, their first battery toy was produced, called "Roller Skating Circus Clown," followed by "Climbing Linesman" in 1957.

The concept of "happy" toys as subject material really stands out in the 1950s. Circus themes are the most predominant in this period, represented by clown skaters, jugglers, cyclists, and musicians. Animals also became a favorite subject and black or African native toys, including "Pango-Pango African Dancer," "Happy Hippo," black "Skating Chef," and "Calypso Joe" were produced.

T.P.S. designs also caught the attention of Louis Marx & Company. Through LINEMAR, the Japanese import arm of Louis Marx, T.P.S. was asked to produce authorized editions of toys based on Popeye and Walt Disney® characters. These toys were usually similar to or based on existing and successful T.P.S. designs. They are easily recognizable as T.P.S. toys because they have the same action or contain common parts. The popular skaters, unicyclists, cyclists, and ball playing toys are illustrations of T.P.S. designs turned into LINEMAR character toys. Another desirable toy, "Juggling Popeye and Olive Oyl," was based on the "Clown Juggler with Monkey." Since LINEMAR functioned only as an importer and had no manufacturing plants in Japan, T.P.S. as well as other companies viewed this as a great opportunity to increase their production while providing toys based on popular and desirable characters.

Clowns on stilts.

Musicians.

Unicyclists in formation.

# 1960S

The early 1960s saw the continuation of the circus and animal themes but with a stronger predominance of animal toy subjects. String climbing toys were introduced in the early 1960s and Tippy Toys™ were introduced in 1965. More toys with amusement park themes called "Dreamland" or "Playland" began to appear. This period also included the introduction of the platform base toys. These toys proliferated during this period with over forty known variations. With beautifully lithographed scenes on the base, they featured various windup vehicles or animals. These toys were so popular that T.P.S. introduced a line of separate miniature vehicles based on the vehicles used for the platform toys. The powerful and well geared windup motor for these vehicles was advertised as able to propel the vehicles for 10 meters (almost 33 feet).

During this period, Japanese character toys became popular, resulting in the release of toys based on Tetsuwan Atom (Astroboy) and his sister Uran, as well as Tetsujin 28 (Gigantor).

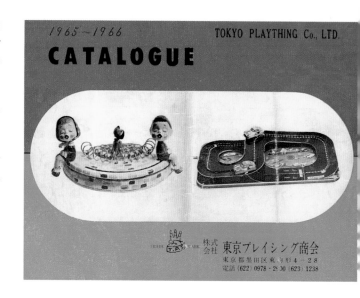

Cover and inside pages from 1965-66 catalog.

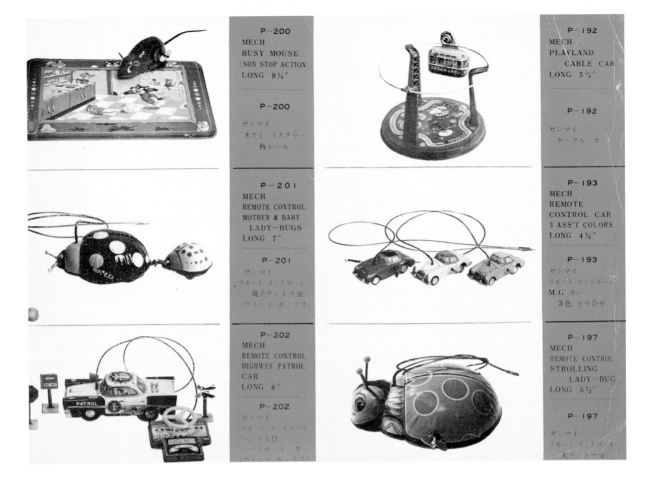

| P-200 | P-192 |
|---|---|
| MECH<br>BUSY MOUSE<br>(NON STOP ACTION)<br>LONG 9¾" | MECH<br>PLAYLAND<br>CABLE CAR<br>LONG 5¾" |
| P-200<br>ゼンマイ<br>ネズミ ミステリー<br>板レール | P-192<br>ゼンマイ<br>ケーブル カー |
| P-201<br>MECH<br>REMOTE CONTROL<br>MOTHER & BABY<br>LADY—BUGS<br>LONG 7" | P-193<br>MECH<br>REMOTE<br>CONTROL CAR<br>3 ASS'T COLORS<br>LONG 4¼" |
| P-201<br>ゼンマイ<br>リモート コントロール<br>親子テントウ虫<br>(ウインド ボックス) | P-193<br>ゼンマイ<br>リモート コントロール<br>M.G カー<br>3色 とり合せ |
| P-202<br>MECH<br>REMOTE CONTROL<br>HIGHWAY PATROL<br>CAR<br>LONG 6" | P-197<br>MECH<br>REMOTE CONTROL<br>STROLLING<br>LADY—BUG<br>LONG 5½" |
| P-202<br>ゼンマイ<br>リモート コントロール<br>ハンドル付<br>パトロール カー<br>(ウインド ボックス) | P-197<br>ゼンマイ<br>リモート コントロール<br>大テントウ虫 |

While T.P.S. had experimented with a battery operated toy in the 1950s, like most Japanese manufacturers they did not begin significant production of battery operated toys until the 1960s. The clever battery powered "Flying Air Car" hovercraft made of lightweight aluminum was introduced in 1966 and became the basis for an Army version and the "Luna Hovercraft." Production of battery powered platform toys with vibrating action to move vehicles along a track began during 1967 and helicopters with very realistic flying action were introduced in 1968.

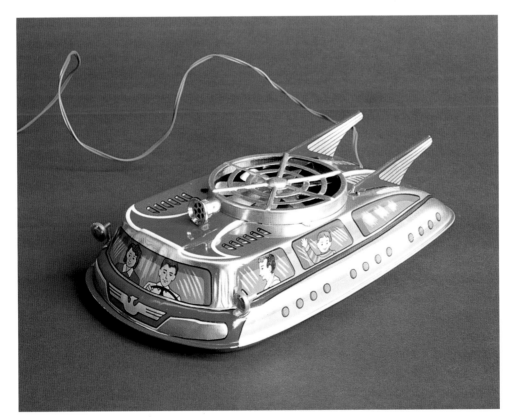

"Flying Air Car" hovercraft.

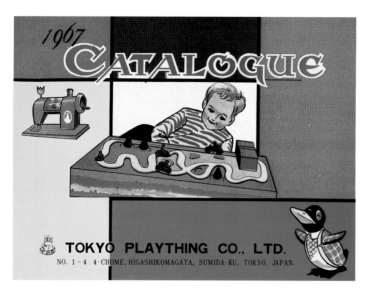

Front and back covers of 1967 catalog.

# 1970S

The world's interest in space exploration led to the popularity of space toys. Although the first space oriented, battery operated toys were produced in the late 1960s, this trend continued into the 1970s. In addition, cars, trucks, and planes became popular. More of the company's production was devoted to battery toys, although miscellaneous mechanical toys continued to be produced as well. Plastic also played a bigger part in toy construction. Toys of mixed construction were evident, as were all-plastic toys by the late 1970s. This change in construction materials did not detract from the creativity at T.P.S., as the actions of these toys continued to be some of the most innovative in the industry.

The popularity of accurate tin car models by notable companies such as Bandai, Ichiko, and Asahi did not go unnoticed. T.P.S. introduced a Ford Mustang based on a 1970 model and also a Porsche 911. Both of these cars served as the basis for variations for more than a decade to follow. These cars would do barrel rolls, screeching turns, complete flip overs, and U-turns, and could travel in forward and reverse directions. These actions continued the tradition of unique, complex, and entertaining toy designs from T.P.S. Motorcycles and planes rounded out the battery powered choices from Toplay. Each of them included some actions not typically found in most toys.

Finally, low cost plastic robots were made during this period. Most were chrome plated, walking, or sparking robots with a good windup motor that allowed long running times.

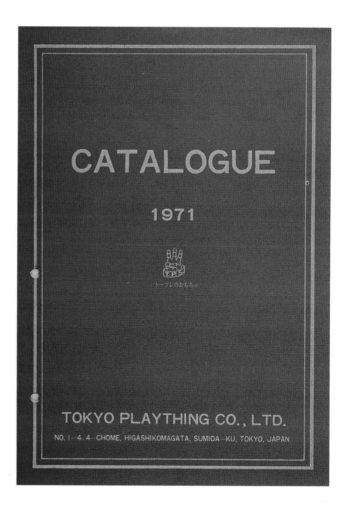

1971 Catalog.

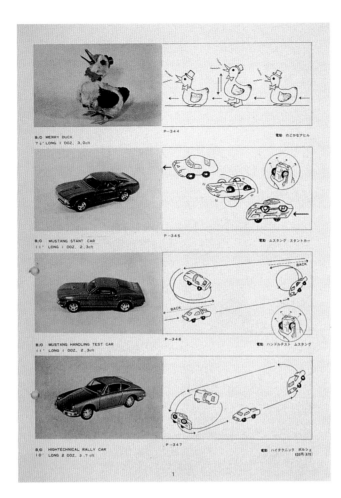

B/O MERRY DUCK
7¾" LONG 1 DOZ. 3.0cft
P-344
電動 のどかなアヒル

B/O MUSTANG STUNT CAR
11" LONG 1 DOZ. 2.3cft
P-345
電動 ムスタング スタントカー

B/O MUSTANG HANDLING TEST CAR
11" LONG 1 DOZ. 2.3cft
P-346
電動 ハンドルテスト ムスタング

B/O HIGHTECHNICAL RALLY CAR
10" LONG 2 DOZ. 3.7 cft
P-347
電動 ハイテクニック ポルシェ
630円(670)

1

B/O WILD WHEELING DUNE BUGGY
11½" LONG 1 DOZ. 3.5cft
P-313
電動 回転サンドバギー
840円(860)

B/O M-A POLICE HELICOPTER
8" LONG 4 DOZ. 2.8cft
P-273
電動 ミステリーヘリコプター

B/O MERCURY EXPLORER
8" LONG 3 DOZ. 5.0cft
P-292
電動 マジックカラー宇宙船
500円(530)

FRIC. EAGLE CRANE TRUCK
長サ 40センチ 780円(870)
P-310
フリクション
イーグルクレーントラック

FRIC. SAND CONVEYOR TRUCK
8" LONG 2 DOZ. 4.4cft
P-320
フリクション
大ベルトコンベアートラック
780円(860)

FRIC. PROPANE TRUCK
長サ 16.5センチ550円(600)
P-309
フリクション
15インチ プロパントラック

FRIC. SAND CONVEYOR TRUCK
8" LONG 2 DOZ. 4.4cft
P-341
フリクション
中ベルトコンベアートラック
530円(570)

P-308
フリクション
幼稚園バス
長サ 24センチ270円(300)

3

MECH. 18 ASSORTED CARS
2½" LONG 24 DOZ. 1.7cft
P-217
ゼンマイ 18点アソートバラ自動車
80円(90)

MECH. RACERS
3-3/4" LONG 24 DOZ. 1.5cft
P-260
MECH. PEANUT RACERS
3½" LONG 24 DOZ. 2.3cft
P-280
ゼンマイ ピーナツレーサー
90円(100)

MECH. DOUBLE-FACED RACE CAR
3-3/4" LONG 12 DOZ. 1.5cft
P-340
MECH. SILVER TANK
3" LONG 12 DOZ. 1.5cft
P-325
ゼンマイ シルバータンク
100円(110)

P-352
ゼンマイ ミサイル戦車
P-353
フリクション ミサイルキャリアー

8

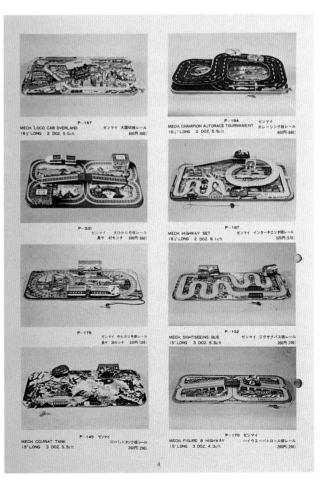

MECH. LOCO CAB OVERLAND
16½" LONG 2 DOZ. 5.0cft
P-167
ゼンマイ 大陸切板レール
600円(660)

MECH. CHAMPION AUTORACE TOURNAMENT
16½" LONG 2 DOZ. 5.5cft
P-194
ゼンマイ 大レーシング板レール
600円(660)

P-331
ゼンマイ 大ひかり号板レール
長サ 42センチ 600円(660)

MECH. HIGHWAY SET
16½" LONG 2 DOZ. 6.1cft
P-187
ゼンマイ インターチェンヂ板レール
520円(570)

P-179
ゼンマイ 中ヒカリ号板レール
長サ 30センチ 260円(300)

MECH. SIGHTSEEING BUS
15" LONG 3 DOZ. 5.5cft
P-132
ゼンマイ ジグザグバス板レール
260円(290)

MECH. COMBAT TANK
15" LONG 3 DOZ. 5.5cft
P-145
ゼンマイ
コンバットタンク板レール
260円(290)

MECH. FIGURE 8 HIGHWAY
15" LONG 3 DOZ. 4.3cft
P-170
ゼンマイ
ハイウエーパトロール板レール
260円(290)

4

# 1980S

While only a few toys from the 1980s have been included in this book as illustrations of the period, this was a time of change. Production costs in Japan were higher than those in developing countries, making local production costly. Large global toy companies were major forces in the market. As a result, many Japanese companies turned to other countries to manufacture their products. Toplay was no exception.

T.P.S. toys produced in Japan were sold in the 1980s, but during this time T.P.S. established licensing arrangements with Taiwanese companies to allow those companies to produce and distribute toys designed by T.P.S. The best example of this relationship is the Dah Yang Toy Industrial Co., Ltd. (D.Y.Toy), which has been producing many of the popular climb and slide battery powered toys found in toy stores today. These are toys in which the subject climbs or is carried to the top before rolling down a slide to the bottom and repeating the action.

1981 Catalog.

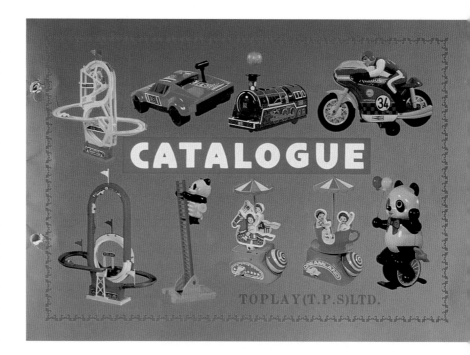

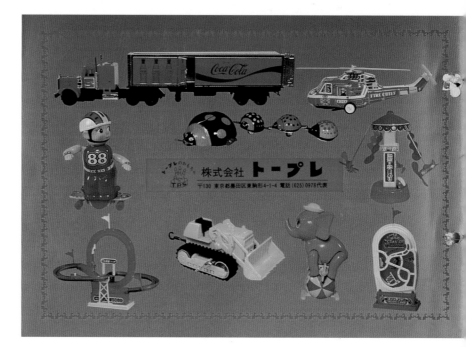

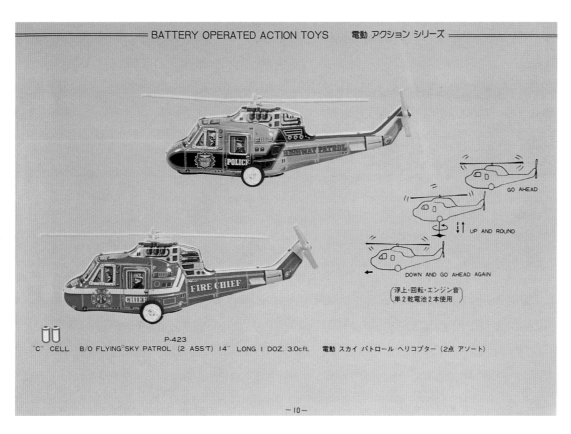

GO AHEAD

UP AND ROUND

DOWN AND GO AHEAD AGAIN

(浮上・回転・エンジン音)
(単 2 乾電池 2 本使用)

P-423
"C" CELL 　B/O FLYING SKY PATROL (2 ASS'T) 14" LONG 1 DOZ. 3.0cft. 　電動 スカイ パトロール ヘリコプター (2点 アソート)

— 10 —

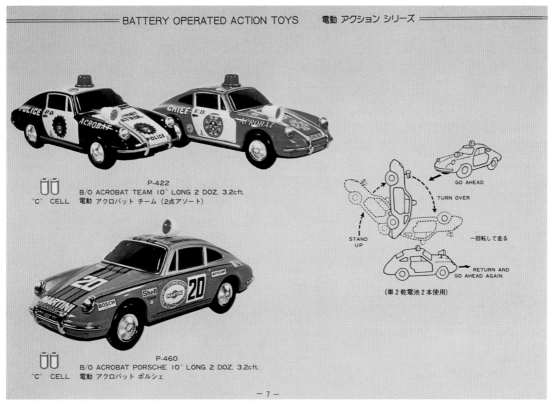

P-422
"C" CELL 　B/O ACROBAT TEAM 10" LONG 2 DOZ. 3.2cft.
電動 アクロバット チーム (2点アソート)

GO AHEAD

TURN OVER

一回転して走る

STAND
UP

RETURN AND
GO AHEAD AGAIN

(単 2 乾電池 2 本使用)

P-460
"C" CELL 　B/O ACROBAT PORSCHE 10" LONG 2 DOZ. 3.2cft.
電動 アクロバット ポルシェ

— 7 —

## 1990S

Toplay (T.P.S.) Ltd. is still in business today. Under the capable leadership of Kazuhiko Udagawa, the company continues to be active in toy design and still licenses those designs to others to manufacture toys. You can find toys from T.P.S. with their own mark made for the Japanese market as well as toys from others selling to the world market under license from T.P.S. The next time you see a clever toy from D.Y.Toy, check the box. You may well find it was made under license from T.P.S. At age seventy-five, Mr. Yoshio Udagawa is still thinking of ideas for toys to make you smile.

Three pages from the 1992 catalog.

Two pages from the 1997 D.Y. Toy catalog.

## Holiday Happiness

**Holiday Express**
Item No.2011J/GB - Pkg: 1 Doz./4.4'

**Cherry Santa**
Item No.2019J/GB - Pkg: 1 Doz./2.5'

**Merry MotorBall**
Item No.8038J/BL - Pkg: 4 Doz./3.3'

**SantaLand**
Item No.2017J/GB - Pkg: 1 Doz./2.7'

**Santa's Castle**
Item No.2045J/GB - Pkg: 1 Doz./2.5'

© 1997 Dah Yang Industrial Co., Ltd.

**D.Y. TOY ®**
Dah Yang Toy Industrial Co., Ltd.
560 Chung Hsiao East Road, Sec. 4, 10th Floor, Taipei, Taiwan
Telephone: 886-2-27582701 Fax: 886-2-27290020
URL: http://www.DYTOY.com E-mail: sales@DYTOY.com

---

## Micro Master Golf

Micro-Master Golf is a self-contained motorized golf game. Try to chip the golf balls into the moving hole on the green. You control the swing with external controls. Ball compartment stores the balls when not in play.

Item No.8081/BL - Pkg: 2 Doz./3.6'

## Micro Dolphin Dunk

Dolphin Dunk is a self-contained motorized ball game. Try to knock the beach balls into the basket as they climb up the steps. You control the dolphin with external controls. Ball compartment stores the beach balls when not in play.

Item No.8082/BL - Pkg: 2 Doz./3.6'

## Micro-League Baseball

Micro-League Baseball is a self-contained motorized baseball game. Try to hit the balls past the outfielder for the home run. You control the swing with external controls. Ball compartment stores the baseballs when not in play.

Item No.8083/BL - Pkg: 2 Doz./3.6'

© 1997 Dah Yang Toy Industrial Co., Ltd.

**D.Y. TOY ®**
Dah Yang Toy Industrial Co., Ltd.
560 Chung Hsiao East Road, Sec. 4, 10th Floor, Taipei, Taiwan
Telephone: 886-2-27582701 Fax: 886-2-27290020
URL: http://www.DYTOY.com E-mail: sales@DYTOY.com

Three pages from the 1998 D.Y.Toy catalog.

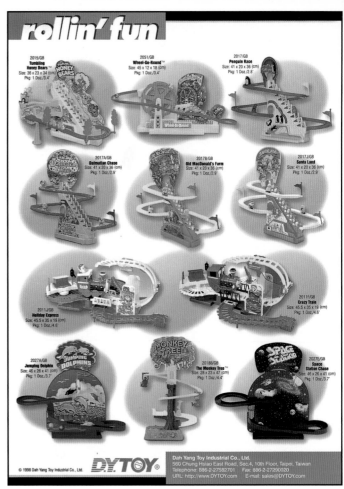

2015/GB
**Tumbling Honey Bears**™
Size: 36 x 23 x 34 (cm)
Pkg: 1 Doz./3.4'

2051/GB
**Wheel-Go-Round**™
Size: 45 x 12 x 18 (cm)
Pkg: 1 Doz./3.4'

2017/GB
**Penguin Race**™
Size: 41 x 20 x 36 (cm)
Pkg: 1 Doz./2.8'

2017A/GB
**Dalmatian Chase**™
Size: 41 x 20 x 36 (cm)
Pkg: 1 Doz./2.9'

2017B/GB
**Old MacDonald's Farm**™
Size: 41 x 20 x 36 (cm)
Pkg: 1 Doz./2.9'

2017J/GB
**Santa Land**™
Size: 41 x 20 x 36 (cm)
Pkg: 1 Doz./2.9'

2011J/GB
**Holiday Express**™
Size: 45.5 x 35 x 19 (cm)
Pkg: 1 Doz./4.6'

2011F/GB
**Crazy Train**™
Size: 45.5 x 35 x 19 (cm)
Pkg: 1 Doz./4.6'

2027A/GB
**Jumping Dolphin**™
Size: 46 x 26 x 41 (cm)
Pkg: 1 Doz./3.7'

2018B/GB
**The Monkey Tree**™
Size: 28 x 23 x 47 (cm)
Pkg: 1 Doz./4.4'

2027E/GB
**Space Station Chase**™
Size: 46 x 26 x 41 (cm)
Pkg: 1 Doz./3.7'

© 1998 Dah Yang Toy Industrial Co., Ltd.

**DY.TOY**®

Dah Yang Toy Industrial Co., Ltd.
560 Chung Hsiao East Road, Sec.4, 10th Floor, Taipei, Taiwan
Telephone: 886-2-27582701    Fax: 886-2-27290020
URL: http://www.DYTOY.com    E-mail: sales@DYTOY.com

# Disney®Fun

**Merry Ball Land**
2048A/GB
Size: 46.5 x 28 x 32 (cm)
Pkg: 1 Doz./2.9'

**The Great Dalmatians Chase**
2017A/GB
Size: 41 x 21.5 x 37 (cm)
Pkg: 1 Doz./2.8'

**Mickey for Kids Playground**
2013/GB
Size: 41 x 21.5 x 37 (cm)
Pkg: 1 Doz./2.8'

**Merry Ball Land**
2048B/GB
Size: 46.5 x 28 x 32 (cm)
Pkg: 1 Doz./2.9'

2011H/GB
Size: 45.5 x 35 x 18.8 (cm)
Pkg: 1 Doz./4.6'

**Mickey for Kids Tumble Train**

*For Disney Licensee Only

© 1998 DISNEY

Dah Yang Toy Industrial Co., Ltd.
10th Floor, No.560, Sec.4, Chung Hsiao East Road, Taipei, Taiwan
Telephone: 886-2-27582701   Fax: 886-2-27290020
URL:http://www.DYTOY.com   E-mail: sales@DYTOY.com

**DY.TOY**®                                    © 1998 Dah Yang Toy Industrial Co., Ltd

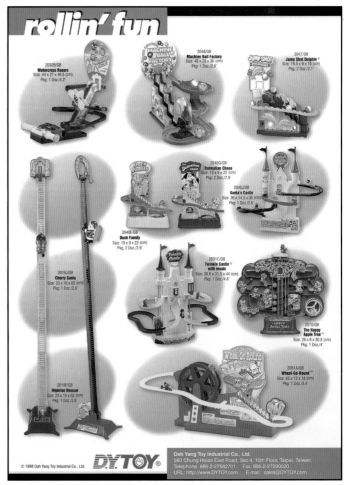

rollin' fun

2032B/GB
**Motorcross Racers**™
Size: 44 x 27 x 46.5 (cm)
Pkg: 1 Doz./4.2'

2048/GB
**Machine Ball Factory**™
Size: 46 x 28 x 30 (cm)
Pkg: 1 Doz./2.9'

2047/GB
**Jump Shot Dolphin**™
Size: 19.5 x 9 x 18 (cm)
Pkg: 2 Doz./2.1'

2040G/GB
**Dalmatian Chase**™
Size: 19 x 9 x 22 (cm)
Pkg: 2 Doz./2.9

2045J/GB
**Santa's Castle**™
Size: 36 x 14.5 x 35 (cm)
Pkg: 1 Doz./2.6'

2040E/GB
**Duck Family**™
Size: 19 x 9 x 22 (cm)
Pkg: 2 Doz./2.9'

2019J/GB
**Cherry Santa**™
Size: 23 x 16 x 62 (cm)
Pkg: 1 Doz./2.6'

2031C/GB
**Twinkle Castle**™ with music
Size: 39.6 x 31.5 x 44 (cm)
Pkg: 1 Doz./4.6'

2070/GB
**The Happy Apple Tree**™
Size: 26 x 8 x 30.5 (cm)
Pkg: 1 Doz./4'

2051A/GB
**Wheel-Go-Round**™
Size: 45 x 12 x 18 (cm)
Pkg: 1 Doz./3.4'

2019F/GB
**Hi prize Rescue**™
Size: 23 x 16 x 62 (cm)
Pkg: 1 Doz./2.6'

© 1998 Dah Yang Toy Industrial Co., Ltd.    **DY.TOY**®

Dah Yang Toy Industrial Co., Ltd.
560 Chung Hsiao East Road, Sec.4, 10th Floor, Taipei, Taiwan
Telephone: 886-2-27582701    Fax: 886-2-27290020
URL: http://www.DYTOY.com    E-mail: sales@DYTOY.com

# MAKERS, MARKETING, AND MARKS

## THE MAKERS

Most early T.P.S. toys were a success, and during the company's peak years employment eventually reached one hundred. But this alone does not indicate the scope of a company's operations. In order to understand the Japanese toy industry, it is necessary to break down the functions involved in making a toy. These include idea, design, engineering, manufacturing, assembly, packaging, shipping, and sales. The use of subcontractors for many of these functions was very common in Japan. This is also one of the reasons it is sometimes difficult to understand the multiple trademarks seen on toys — but more about that later.

Companies such as Toplay, who were involved in determining what products to bring to the market, usually had creative people with toy ideas. In the case of Toplay, this role was filled primarily by Mr. Udagawa. A designer would take these ideas and convert them to outline and artwork drawings. An engineer then made these designs work and created the drawings necessary for manufacturing and assembly.

It was then common for a subcontractor to print and stamp the tin, or stamp and paint the tin before sending these parts to the assembling factory. Other suppliers would provide motors, gears, rubber wheels and parts, plastic parts, etc. to the final assembly location. The subcontract concept meant that a toy company in Japan did not necessarily have to have in-house capability for all of these functions. As a result, there were literally hundreds of manufacturers and thousands of subcontractors.

T.P.S. would also use subcontractors for parts and for complete toys. Often the subcontractor's mark was included on the toy. Subcontractors whose marks can be found on some T.P.S. toys include: Hikari, Hirata, and Hishimo. Toplay would also function as a subcontractor to others as a design or manufacturing source. LINEMAR and Mitsuhashi are two examples of where Toplay served as a subcontractor to other companies.

Hikari trademark.

Hirata trademark.

Hishimo trademark.

## MARKETING

Japanese companies typically used trading companies to access the major export markets outside of Japan, such as North America and Europe. There were hundreds of trading companies located in Tokyo, Osaka, and Kobe, and many of them represented multiple manufacturers. These trading companies would in turn sell to foreign buyers or importers whose names also often appeared on the toy or on the box. These buyers would come to visit the trading companies, who in turn would have samples on display from many toy competitors like Toplay, Tomiyama, Masudaya, Bandai, and Nomura.

The importers then marketed these toys in their respective markets. While maybe 25-30% of T.P.S. toys were exported to Europe, the majority were exported to North America. Names such as Cragstan and LINEMAR are prime examples of importers. These companies imported toys to North America and did not have Japanese manufacturing plants, yet their names were often found on the toy and/or on the box. The trading companies were the sellers and the importers were the buyers. The trading companies would try to sell the products from the manufacturers and the importers would tell the manufacturers, through the trading companies, what toys they wanted them to build. Accordingly, both played an important role in the marketing and sales of toys from Japan.

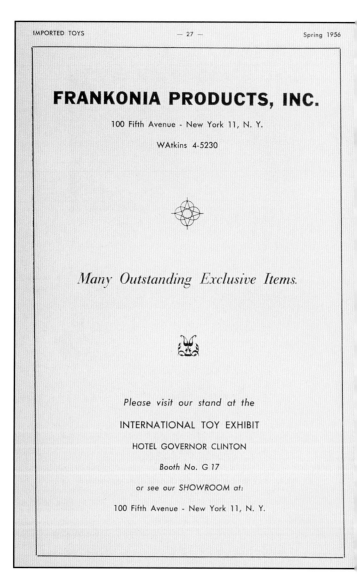

Cragstan ad from 1956 International Toy Exhibit.

The following importers names can be found on T.P.S. boxes or toys:

Cragstan (Craig - Stanton - Elmaleh, Incorporated of New York)
A.H.I. (Azrak - Hamway, Inc., New York)
Franconia (Franconia Products, Inc., New York)
ROSKO
Shackman (B. Shackman & Company, New York)
Mego (Mego Corporation, New York)
LINEMAR (Louis Marx line importing office in Japan)
Sonsco
F.E. White (F.E.White Co., Inc., New York)
Leadworks (Leadworks, Cleveland, Ohio)

The following chart illustrates this process using T.P.S. as the example:

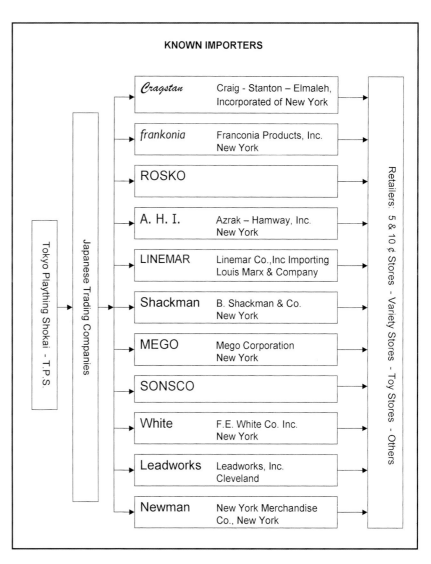

A.H.I. ad from 1956 International Toy Exhibit.

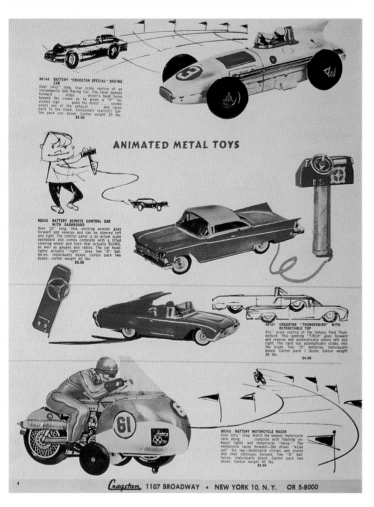

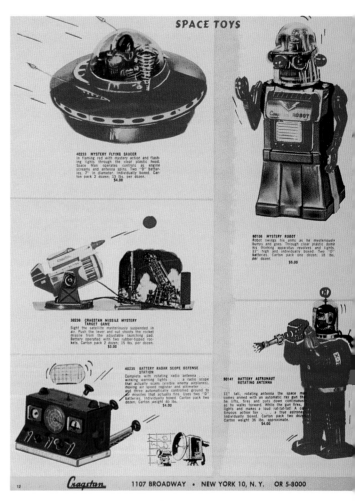

Five pages from the 1961 Cragstan catalog.

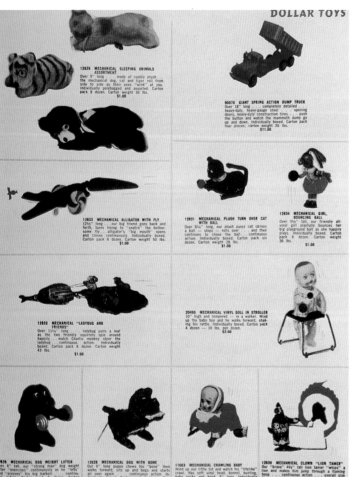

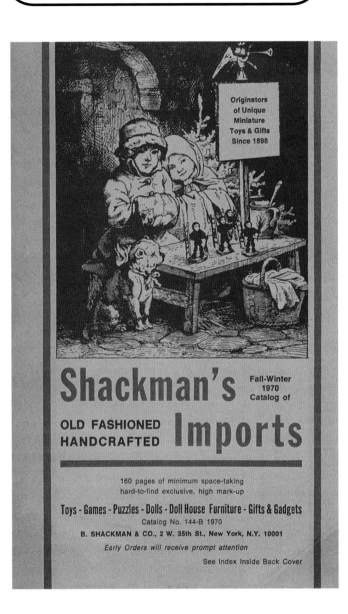

Below and next page:
Cover and two pages from the 1970 Shackman catalog.

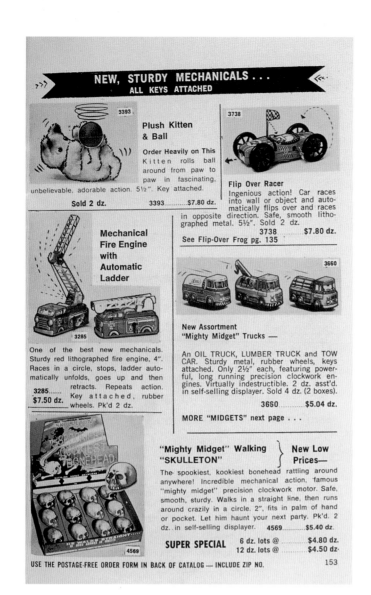

# MARKS

You can begin to see the number of companies involved in the process of bringing a toy to market. Since these companies may or may not add their mark to the toy or box, it is no wonder that we find Japanese trademarks very confusing. As we try to understand who designed, manufactured, sold, or imported a toy, it is often an impossible task. We may see one, two, or three trademarks on the toy or box. The trademarks could belong to the lead toy company, a subcontractor, a trading company, an importer, or any combination of these.

While toy companies sought their own identity, they were primarily focused on production volume instead of creating brand preference. In fact, the original brands were often hidden just because the importer was promoting its own brand. Therefore, time has done more to raise the awareness of brands than did intent. As collectors have been attracted to certain types of toys, they have learned that these toys often came from the same company, causing the collectors to now be aware of the brand and begin

to search for it. Such is the case for T.P.S. toys. They have survived the test of time and have become desirable for their designs, not because we knew who they were in the 1950s.

A word about copies. Copying was a problem then as it is today. Toy companies would often get ideas from each other or try to duplicate successful designs. You can also see the roots of many Japanese toys in designs from Germany or United States. Sometimes the importers would take the designs of one company to another company or even another country, to see if it could be made cheaper. This eventually led to copies of Japanese toys being produced in places like Korea, Hong Kong, Taiwan, and China.

Often "Patent Pending" was written on the box or toy. This was done both to discourage copies as well as to indicate a legitimate patent application in process. If you have tried to research patent numbers taken from toys or boxes, you know they don't match up to the number sequencing from the US Patent Office. This is because the patent numbers referenced were usually for Japanese patents, not US patents.

# TRADEMARK IDENTIFICATION

To help you sort through the many trademarks or brands seen on Japanese toys of this period, the following pages contain over two hundred Japanese trademarks known to have been used, along with the Japanese company name and the adopted English name or the English translation (translations shown in parenthesis). Where possible, the derivation of the mark is indicated. Remember that Japanese is read right to left and a few of the trademark initials reflect this practice; i.e., T.N. stood for Nomura Toy, read right to left. For easier identification, this listing is alphabetical (by the initials used in the trademark), with those companies using symbols listed at the end.

Many companies changed or modified their trademarks during their existence. These changes were sometimes subtle, but occasionally they were more substantial as the company worked at building a brand image. If a mark is similar to one shown in this section, it is possible that it comes from the same company.

Below and on the next page are four examples of well known companies that changed their trademark over the years.

## Asahi

1948-1955

1955-

## Bandai

1950-1961

1961- ? (added Bandai Baby)

Contemporary

## Tomy

1930-1959

1959-1963

1963-1982

1982-

**Yoneya**

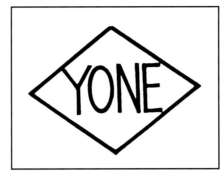

1950-1964

1964-

# TRADEMARK LISTING

There are a few Japanese word translations to keep in mind when reading a company name that will help you to better understand the company's role within the toy industry:

*Gangu* = toy
*Kinzoku* = metal
*Kogyo* = industry
*Sangyo* = industry
*Seisakusho* = manufacturing company or factory
*Shoji* = trading company
*Shokai* = company or merchant
*Shoten* = company or shop

Initials: 5 circle
Company: **Marugo** Shoten
English name: (Marugo Company)
Trademark derivation: Maru go means
**circle + 5**

Initials: A1 + Aksakusa
Company: **Asakusa** Toy Ltd.
English name: Asakusa Toy Ltd.
Trademark derivation: <u>Asakusa</u> Toy Ltd.

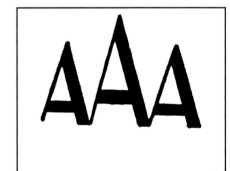

Initials: AAA
Company: **Sanei** Gangu Co., Ltd.
English name: (Sanei Toy Company)
Trademark derivation: Sanei means (3 As)
**AAA**

Initials: Alps
Company: **Alps** Shoji Co. Ltd.
English name: Alps Toy
Trademark derivation: **Alps** Shoji

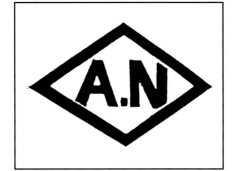

Initials: A.N
Company: **Naito** Shoten
English name: Naito Toy Co., Ltd.
Trademark derivation: Mr. **A**. **N**aito

Initials: AOKI
Company: **Aoki** Shokai
English name: Aoki Ind., Ltd.
Trademark derivation: **Aoki** Shokai

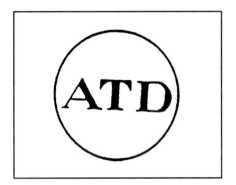

Initials: ASAHITOY (used 1948-1955)
Company: **Asahi** Gangu Seisakusho
English name: Asahi Toy Company, Ltd.
Trademark derivation: **Asahi Toy** Company

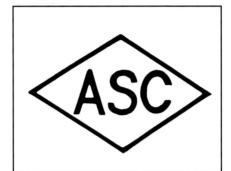

Initials: ASC
Company: **Aoshin** Shoten
English name: Aoshin Shoten Co., Ltd.
Trademark derivation: **A**oshin **S**hoten **C**ompany

Initials: ATC (used after 1955)
Company: **Asahi** Gangu Seisakusho
English name: Asahi Toy Company, Ltd.
Trademark derivation: **A**sahi **T**oy **C**ompany, Ltd.

Initials: ATD
Company: **Asakusa** Toys & Dolls Co., Ltd.
English name: Asakusa Toys & Dolls Co., Ltd.
Trademark derivation: **A**sakusa **T**oys & **D**olls

Initials: B (used 1950-1961)
Company: **Bandaiya**
English name: Bandai Co., Ltd
Trademark derivation: **B**andaiya

Initials: B + Bandai Baby (added 1961)
Company: **Bandai**
English name: Bandai Co., Ltd
Trademark derivation: **B**andai Co., Ltd

Initials: BANDAI
Company: **Bandai** - Present
English name: Bandai Co., Ltd
Trademark derivation: **Bandai** Co., Ltd

Initials: Billiken
Company: **Billiken** Shokai
English name: Billiken Co., Ltd.
Trademark derivation: **Billiken** Shokai

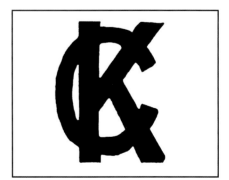

Initials: CK
Company: **Koshibe** Shoten
English name: (Koshibe Company)
Trademark derivation: Mr. **C**. **K**oshibe

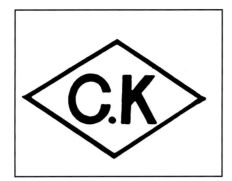

Initials: CK
Company: **Kuramochi** Shoten
English name: Kuramochi & Co., Ltd.
Trademark derivation: <u>K</u>uramochi <u>C</u>o.

Initials: DAITO
Company: **Daito** Co., Ltd.
English name: Daito Co., Ltd.
Trademark derivation: <u>Daito</u> Co., Ltd.

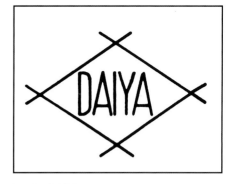

Initials: DAIYA
Company: **Terai** Shoten
English name: Terai Toys Co., Ltd. (Mr. Terai)
Trademark derivation: <u>Daiya means diamond</u>

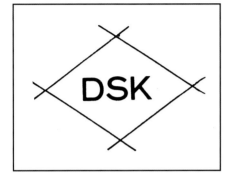

Initials: DSK
Company: **Daishin** Kogyo
English name: (Daishin Industry)
Trademark derivation: <u>Dai</u>shin <u>K</u>ogyo

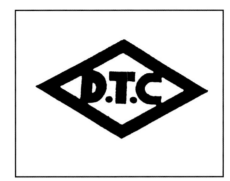

Initials: DTC
Company: **Daiwa** Toys Co., Ltd.
English name: Daiwa Toys Co., Ltd.
Trademark derivation: <u>D</u>aiwa <u>T</u>oys <u>C</u>o., Ltd.

Initials: Epoch + Raccoon
Company: **Epoch** Co., Ltd
English name: Epoch Co., Ltd
Trademark derivation: **Epoch** Co., Ltd

Initials: E.T.Co. (used 1930-1959)
Company: **Tomiyama** Shoji
English name: Tomiyama Trading Co., Ltd.
Trademark derivation: <u>E</u>iichiro <u>T</u>omiyama
<u>Co</u>mpany

Initials: Fujitaka
Company: **Fujitaka** Co. Ltd.
English name: Fujitaka Co. Ltd.
Trademark derivation: **Fujitaka**

Initials: H
Company: **Hasegawa** Gangu Seisakusho
English name: (Hasegawa Toy Manufacturing)
Trademark derivation: <u>H</u>asegawa

Initials: H
Company: **Hashimoto** Co., Ltd.
English name: Hashimoto Co., Ltd.
Trademark derivation: <u>H</u>ashimoto

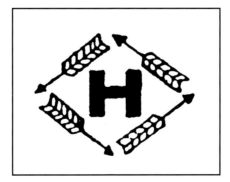

Initials: H
Company: **Hayashi** Seisakusho
English name: (Hayashi Manufacturing)
Trademark derivation: <u>H</u>ayashi

Initials: H
Company: **Hirata** Gangu Seisakusho
English name: (Hirata Toy Manufacturing)
Trademark derivation: <u>H</u>irata

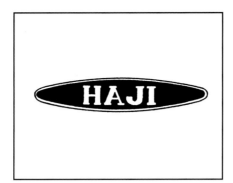

Initials: H
Company: **Hitachi** Gangu Seisakusho
English name: (Hitachi Toy Manufacturing)
Trademark derivation: **H**itachi

Initials: HAJI
Company: **Mansei** Gangu
English name: Mansei Toys Co., Ltd
Trademark derivation: Mr. Y. **Haji** (Owner)

Initials: HIRO
Company: Unknown
English name:
Trademark derivation:

Initials: HK
Company: **Kitazawa** & Co., Ltd
English name: Kitazawa & Co., Ltd
Trademark derivation: Mr. **H**. **K**itazawa

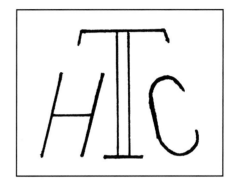

Initials: H.S.
Company: **Harusame** Seisakusho
English name: (Harusame Manufacturing)
Trademark derivation: **H**arusame **S**eisakusho

Initials: HTC
Company: Unknown
English name:
Trademark derivation:

Initials: HTP
Company: **Hikari** Gangu Seisakusho
English name: (Hikari Toy Manufacturing)
Trademark derivation: **H**ikari **T**oy **P**roducts

Initials: I + circle
Company: **Ichimura** Seisakusho
English name: (Ichimura Manufacturing)
Trademark derivation: **I**chimura

Initials: ICHIKO + PU
Company: **Ichiko** Kogyo
English name: Ichico Manufacturing Co., Ltd
Trademark derivation: **Ichico** Manufacturing
Co., Ltd

Initials: IK
Company: **Ichiyo** Kogyo
English name: (Ichiyo Industry)
Trademark derivation: **I**chiyo **K**ogyo

Initials: IMAI + H
Company: **Imai** Kogyo
English name: (Imai Industry)
Trademark derivation: **Imai** Kogyo

Initials: Inakita
Company: **Inakita** Shokai
English name: Inakita Co., Ltd
Trademark derivation: **Inakita** Shokai

Initials: IS
Company: **Imakita** Kinzoku Kogyo
English name: (Imakita Metal Manufacturing)
Trademark derivation: Mr. **S**. **I**makita

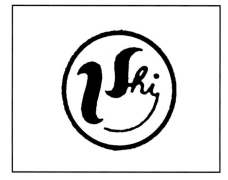

Initials: ISHI
Company: **Ishizuka** Seisakusho
English name: (Ishizuka Manufacturing)
Trademark derivation: **Ishi**zuka

Initials: ITS
Company: **Ishizaki** Gangu Seisakusho
English name: (Ishizaki Toy Manufacturing)
Trademark derivation: **I**shizaki **T**oy **S**eisakusho

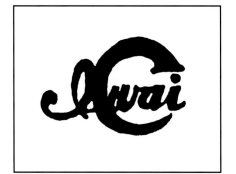

Initials: IWAI
Company: **Iwai** Sangyo Ltd.
English name: (Iwai Ind., Ltd)
Trademark derivation: **Iwai** Sangyo Ltd.

Initials: K
Company: **Kashiwai** Seisakusho
English name: Kashiwai Metal Works Ltd.
Trademark derivation: **K**ashiwai

Initials: K
Company: **Kawahara** Gangu Seisakusho
English name: (Kawahara Toy Manufacturing)
Trademark derivation: **K**awahara

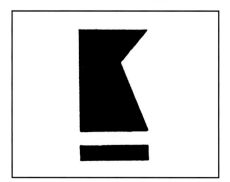

Initials: K
Company: **Keiichi** Kogyosho Co., Ltd.
English name: Keiichi Kogyosho Co., Ltd.
Trademark derivation: **K**eiichi Kogyosho

Initials: K
Company: **Kokyu** Shokai
English name: Kokyu Trading Co., Ltd.
Trademark derivation: **K**okyu Shoklai

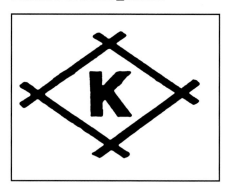

Initials: K
Company: **Koyo** Kinzoku Co., Ltd.
English name: Koyo Metal Co.
Trademark derivation: **K**oyo Kinzoku

Initials: K
Company: **Kusama** Shoten
English name: (Kusama Company)
Trademark derivation: **K**usama

Initials: K
Company: Mitake Gangu Seisakusho
English name: (Mitake Toy Manufacturing)
Trademark derivation: Unknown

Initials: K (KO)
Company: **Ohki** Gangu Seisakusho
English name: (Ohki Toy Manufacturing)
Trademark derivation: Mr. **K**. **O**hki

Initials: K (KO)
Company: **Ohta** Kasaburo Co., Ltd.
English name: Ohta Kasaburo Co., Ltd.
Trademark derivation: Mr. **K**asaburo **O**hta

Initials: K (KO)
Company: **Okada** Gangu Seisakusho
English name: (Okada Toy Manufacturing)
Trademark derivation: Mr. **K**. **O**kada

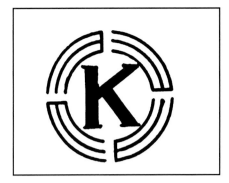

Initials: K
Company: **Sankei** Kogyo
English name: (Sankei Industry)
Trademark derivation: Unknown

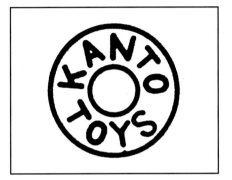

Initials: KANTO TOYS
Company: **Kanto** Toys Co., Ltd.
English name: Kanto Toys Co., Ltd
Trademark derivation: **Kanto** Toys Co., Ltd.

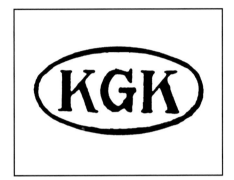

Initials: KGK
Company: **Kyokoku** Gangu Kogyo
English name: (Kyokoku Toy Industry)
Trademark derivation: **K**yokoku **G**angu **K**ogyo

Initials: KHT
Company: **Kawahachi** Shoten
English name: Kawahachi Toy Co., Ltd.
Trademark derivation: **K**awa**H**achi **T**oy

Initials: KIS
Company: **Koiso** Seisakusho
English name: (Koiso Manufacturing)
Trademark derivation: **K**oiso **S**eisakusho

Initials: Kiyoshi
Company: **Kioyshi** Kogei Co. Ltd.
English name: Kioyshi Kogei Co. Ltd.
Trademark derivation: **Kiyoshi**

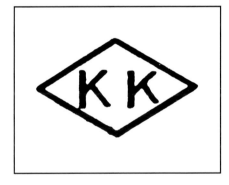

Initials: KK
Company: **Kamiya** Seisakusho
English name: (Kamiya Manufacturing)
Trademark derivation: Mr. **K**. **K**amiya

Initials: KK
Company: **Kobayashi** Seisakusho
English name: (Kobayashi Manufacturing)
Trademark derivation: Mr. **K**. **K**obayashi

Initials: KK
Company: **Kohno**, Kakuzo
English name: Kohno, Kakuzo
Trademark derivation: Mr. **K**akuzo **K**ohno

Initials: KK
Company: **Kowa** Kogyo
English name: (Kowa Manufacturing)
Trademark derivation: **K**owa **K**ogyo

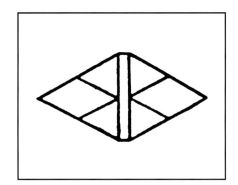

Initials: KK
Company: **Kuwamura** Seisakusho
English name: (Kuwamura Manufacturing)
Trademark derivation: (Mr.) **K**. **K**uwamura

Initials: KKS
Company: **Komoda** Shoten
English name: Komoda Shoten Ltd.
Trademark derivation: (Mr.) **K**. **K**omoda **S**hoten

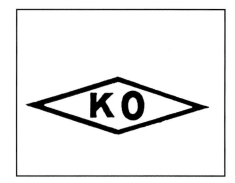

Initials: K.O.
Company: **Yoshiya**
English name: Yoshiya
Trademark derivation: Mr. **K**. **O**hkubo

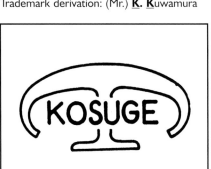

Initials: KOSUGE
Company: **Kosuge** Gangu Kenkyusho
English name: Kosuge Toy Co., Ltd.
Trademark derivation: Matsuzo **Kosuge**

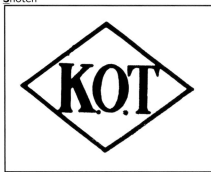

Initials: K.O.T
Company: **Yoshiya**
English name: Yoshiya
Trademark derivation: Mr. **K**. **O**hkubo **T**oy

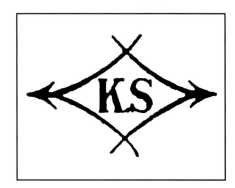

Initials: KS
Company: **Kawano** Seisakusho
English name: (Kawano Manufacturing)
Trademark derivation: **K**awano **S**eisakusho

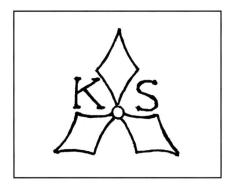

Initials: KS
Company: Unknown
English name:
Trademark derivation:

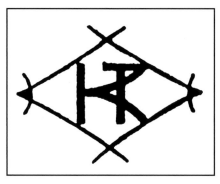

Initials: KT
Company: **Komatsudo** Seisakusho
English name: (Komatsudo Manufacturing)
Trademark derivation: **K**omatsudo **T**oy

Initials: K.Y
Company: **Yagikei** Co., Ltd.
English name: Yagikei Co., Ltd.
Trademark derivation: Mr. **K**. **Y**agi

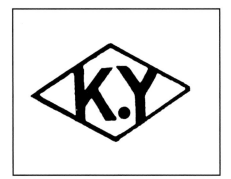

Initials: KY
Company: **Yamazaki** Gomu Gangu Seizo Co.
English name: Yamazaki Gomu Gangu Seizo Co.
Trademark derivation: Mr. **K**. **Y**amazaki

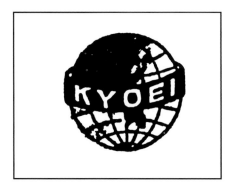

Initials: KYOEI
Company: **Kyoei** Gangu Co., Ltd.
English name: Kyoei Toy Co., Ltd.
Trademark derivation: **Kyoei**

Initials: M
Company: **Mikuni** Sangyo
English name: (Mikuni Industry)
Trademark derivation: **M**ikuni

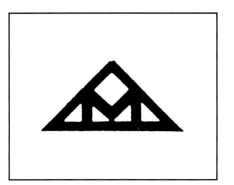

Initials: M
Company: **Mitsuhashi**
English name: Mitsuhashi & Co., Ltd
Trademark derivation: Mr. I. **M**itsuhashi

Initials: M
Company: **Mitsuwa** Gangu Co., Ltd.
English name: (Mitsuwa Toy Co.)
Trademark derivation: **M**itsuwa

Initials: M
Company: **Miura** Shoji Co., Ltd.
English name: Miura Toy Co.
Trademark derivation: Mr. K. **M**iura

Initials: M
Company: **Shimizu** Gangu
English name: (Shimizu Toy)
Trademark derivation: Unknown

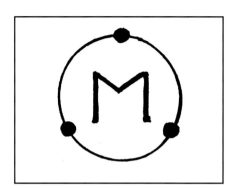

Initials: M
Company: Unknown
English name:
Trademark derivation:

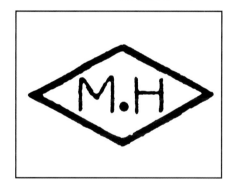

Initials: MARUBISHI
Company: **Marubishi** Co., Ltd
English name: Marubishi Co., Ltd
Trademark derivation: **Marubishi** Co., Ltd

Initials: MEIHO
Company: **Meiho** Shoji
English name: Meiho Trading Co.
Trademark derivation: **Meiho**

Initials: MEIKO
Company: **Meiko** Shoji
English name: (Meiko Trading Co.)
Trademark derivation: **Meiko** Shoji

Initials: M.H
Company: **Hasebe** Yushutsu Gangu
Seisakusho
English name: (Hasebe Export Toy Mfg.)
Trademark derivation: Mr. **M**. **H**asebe

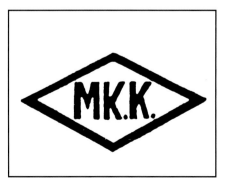

Initials: MKK
Company: **Meiwa** Kogyo
English name: (Meiwa Industry)
Trademark derivation: **M**eiwa **K**ogyo

Initials: MM
Company: **Mitsushima** Seisakusho
English name: (Mitsushima Manufacturing)
Trademark derivation: Mr. **M**. **M**itsushima

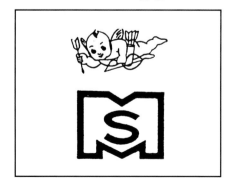

Initials: MS
Company: **Masuya** Gangu
English name: Masuya Toys Co., Ltd.
Trademark derivation: **M**asuya (Mr. K. **S**hizuta)

33

Initials: MT
Company: **Maruyoshi** Gangu Seisakusho
English name: Maruyoshi Toy Mfg. Co., Ltd
Trademark derivation: Mr. K. **M**aruyoshi **T**oy

Initials: MT
Company: **Masudaya** Saito Bokei
English name: Masudaya Toys Co., Ltd.
Trademark derivation: **M**asudaya **T**oys Co.,
Ltd. Modern Toys (Laboratory) is part of
trademark, but not company name.

Initials: M.T.H
Company: **Hasegawa** Shoten
English name: Hasegawa Toy Mfg.
Trademark derivation: **H**asegawa **T**oy
**M**anufacturing

Initials: MTS
Company: **Mutsu** Seisakusho
English name: (Mutsu Manufacturing)
Trademark derivation: **M**utsu **T**oy **S**eisakusho

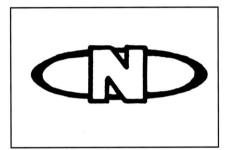

Initials: N
Company: **Nakayama** Shokai
English name: (Nakayama Company)
Trademark derivation: **N**akayama

Initials: N
Company: **Nihon** Gangu Co., Ltd.
English name: Nihon Gangu Co., Ltd.
Trademark derivation: **N**ihon Gangu

Initials: N
Company: **Nippon** Kogei Co., Ltd.
English name: Nippon Kogei Co., Ltd.
Trademark derivation: **N**ippon

Initials: N
Company: **Noguchi** Shoten Co. Ltd.
English name: Noguchi Shoten Co. Ltd.
Trademark derivation: **N**oguchi

Initials: N
Company: Unknown
English name:
Trademark derivation:

Initials: N + turtle
Company: **Nakajima** Seisakusho
English name: (Nakajima Manufacturing)
Trademark derivation: **N**akajima

Initials: Nemoto
Company: **Nemoto** Shoten Co., Ltd.
English name: Nemoto Shoten Co., Ltd.
Trademark derivation: **Nemoto** Shoten Co.,
Ltd.

Initials: NGS
Company: Unknown
English name:
Trademark derivation:

Initials: N.S
Company: **Noguchi** Seisakusho
English name: (Noguchi Manufacturing)
Trademark derivation: **N**oguchi **S**eisakusho

Initials: NT
Company: **Nakamura** Gangu Seisakusho
English name: (Nakamura Toy Manufacturing)
Trademark derivation: **N**akamura **T**oy

Initials: NT
Company: **Nawata** Shoten Co., Ltd
English name: Nawata Shoten Co., Ltd
Trademark derivation: Mr. **T**. **N**awata

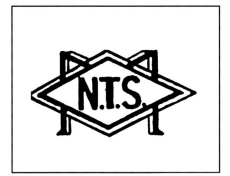

Initials: NTS
Company: **Nakada** Sangyo
English name: (Nakada Industry)
Trademark derivation: **N**akada **T**oy **S**angyo

Initials: NY
Company: **Yoshida** Gangu Seisakusho
English name: (Yoshida Toy Manufacturing)
Trademark derivation: Mr. **N**obuharu **Y**oshida

Initials: OK
Company: K. **Odagawa** Shoten Co., Ltd.
English name: K. Odagawa Co., Ltd.
Trademark derivation: **K**. **O**dagawa

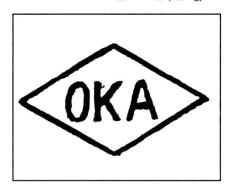

Initials: OKA
Company: **Oka** Gangu Seisakusho
English name: (Oka Toy Manufacturing)
Trademark derivation: **Oka** Gangu

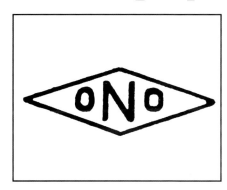

Initials: ONO
Company: **Ono** Toy Co., Ltd
English name: Ono Toy Co., Ltd
Trademark derivation: **Ono** Toy Co., Ltd

Initials: OS
Company: **Oshika** Seisakusho
English name: (Oshika Manufacturing)
Trademark derivation: **O**shika **S**eisakusho

Initials: Plaything
Company: Unknown
English name:
Trademark derivation:

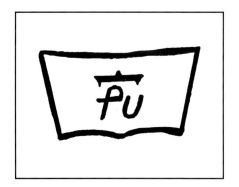

Initials: PU
Company: **Ichiko** Kogyo
English name: Ichico Manufacturing Co., Ltd
Trademark derivation: Unknown

Initials: R.F.
Company: **Fukuda** Seisakusho
English name: (Fukuda Manufacturing)
Trademark derivation: Mr. **R**. **F**ukuda

Initials: RVT
Company: **Iwaya** Seisakusho
English name: Iwaya Corporation (a.k.a. Rock Valley Toys)
Trademark derivation: Iwaya means **R**ock **V**alley **T**oy

Initials: S (SH)
Company: **Harina** Gangu Ltd.
English name: (Harina Toy)
Trademark derivation: Mr. **S**. **H**arina

Initials: S
Company: **Koa** Kinzoku
English name: (Koa Metal)
Trademark derivation: Mr. **S**uzuki

Initials: S
Company: **Sanyo** Shoten Co., Ltd.
English name: Sanyo Toys Co., Ltd.
Trademark derivation: **S**anyo

Initials: S
Company: **Seki** Seisakusho
English name: (Seki Manufacturing)
Trademark derivation: **S**eki

Initials: S
Company: **Shibuya** Gangu Seisakusho
English name: (Shibuya Toy Manufacturing)
Trademark derivation: **S**hibuya

Initials: S & E
Company: **Suzuki and Edwards** Co., Ltd.
English name: Suzuki and Edwards Co., Ltd.
Trademark derivation: **S**uzuki & **E**dwards

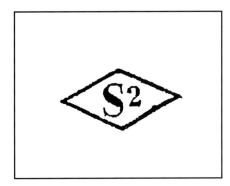

Initials: S2
Company: **Saito** Gangu Co., Ltd.
English name: (Saito Toy Co.)
Trademark derivation: Mr. **S**. **S**aito (**S2**)

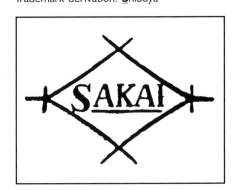

Initials: SAKAI
Company: **Sakai** Seisakusho
English name: Sakai Manufacturing
Trademark derivation: **Sakai**

Initials: SAN circle
Company: **Marusan** Shoten
English name: Marusan Shoten, Ltd.
Trademark derivation: Maru san means **circle + SAN**

Initials: SATO
Company: **Sato** Gangu
English name: (Sato Toy)
Trademark derivation: **Sato**

Initials: SF
Company: **Fumita** Shoten Co., Ltd.
English name: Fumita Shoten Co., Ltd.
Trademark derivation: Mr. **S**. **F**umita

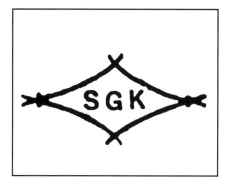

Initials: SGK
Company: **Shimizu** Gangu Kogyo
English name: (Shimizu Toy Industry)
Trademark derivation: **S**himizu **G**angu **K**ogyo

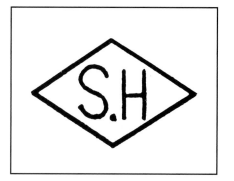

Initials: S.H
Company: **Horikowa** Gangu Kogyo
English name: Horikowa Toy Industrial Co., Ltd.
Trademark derivation: Mr. **S**. **H**orikawa

Initials: SHOWA
Company: **Showa** Kogyo Co., Ltd
English name: (Showa Industry Co.)
Trademark derivation: **Showa**

Initials: SHUDO
Company: **Shudo** Shoji
English name: (Shudo Trading Company)
Trademark derivation: **Shudo**

Initials: SK
Company: **Shinto** Kinzoku Kogyo
English name: Shinto Metal Industrial Ltd.
Trademark derivation: **S**hinto **K**inzoku

Initials: SK
Company: **Suda** Kinzoku Seisakusho
English name: (Suda Metal Manufacturing)
Trademark derivation: **S**uda **K**inzoku

Initials: SKK
Company: **Sinsei** Kiki Kogyo
English name: (Sinsei Toys Industrial Co., Ltd.)
Trademark derivation: **S**insei **K**iki **K**ogyo

Initials: SKK
Company: **Shinsei** Kogyo
English name: (Shinsei Industry)
Trademark derivation: **S**hinsei **K**ogyo **K**abash

Initials: SKT
Company: **Sagane** Gangu Seisakusho
English name: (Sagane Toy Manufacturing)
Trademark derivation: Unknown

Initials: SKT
Company: **Sakin** Gangu Seisakusho
English name: (Sakin Toy Manufacturing)
Trademark derivation: Unknown

Initials: SS
Company: **Shimazaki**
English name: Shimazaki
Trademark derivation: Unknown

Initials: SSS
Company: **Shioji** Shoten
English name: Shioji & Co., Ltd.
Trademark derivation: (Mr.) **S**higeo **S**hioji **S**hoten

Initials: SSS
Company: **Shioji** Shoten
English name: Shioji & Co., Ltd.
Trademark derivation: (Mr.) **S**higeo **S**hioji
**S**hoten

Initials: SSS
Company: **Shioji** Shoten
English name: Shioji & Co., Ltd.
Trademark derivation: (Mr.) **S**higeo **S**hioji
**S**hoten

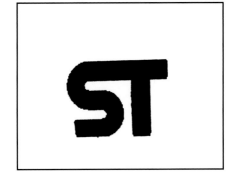

Initials: ST
Not a company trademark, but a **Safe Toy**
designation. Adopted October 1971 by the
Japanese Toy Association as a symbol for
meeting safety standards set by the industry.
Derivation: **S**afe **T**oy

Initials: S.T.S
Company: **Tahara** Seisakusho
English name: (Tahara Manufacturing)
Trademark derivation: (Mr.) **S**. **T**ahara
**S**eisakusho

Initials: S.T.S
Company: **Tamiya** Seisakusho
English name: (Tamiya Manufacturing)
Trademark derivation: **S**. **T**amiya **S**eisakusho

Initials: SUZUKI
Company: **Suzuki** Gangu Seisakusho
English name: (Suzuki Toy Manufacturing)
Trademark derivation: **Suzuki**

Initials: SY
Company: **Sano** Kinzoku Seisakusho
English name: (Sano Metal Manufacturing)
Trademark derivation: Mr. **S**. **Y**amazaki

Initials: SY (used 1950-1964)
Company: **Yoneya**
English name: Yoneya
Trademark derivation: **S**. **Y**oneya

Initials: SZK
Company: **Suzuki** Gangu
English name: (Suzuki Toy)
Trademark derivation: **S**u**Z**u**K**i

Initials: T
Company: **Taiyo** Shokai
English name: (Taiyo Industry)
Trademark derivation: **T**aiyo

Initials: T + Bell
Company: **Tsukuda** Co., Ltd.
English name: Tsukuda Co., Ltd.
Trademark derivation: **T**sukuda

Initials: TADA
Company: **Tada** Seisakusho
English name: (Tada Manufacturing)
Trademark derivation: **Tada**

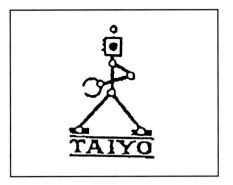

Initials: TAIYO
Company: **Taiyo** Shokai
English name: (Taiyo Industry)
Trademark derivation: **Taiyo**

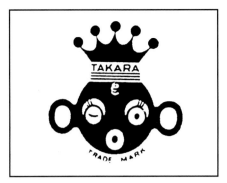

Initials: Takara
Company: **Takara** Vinyl Mfg. Co., Ltd
English name: Takara Vinyl Mfg. Co., Ltd
Trademark derivation: **Takara**

Initials: T.K.
Company: **Higashi** Toy Ltd.
English name: Higashi Toy Ltd.
Trademark derivation: Unknown

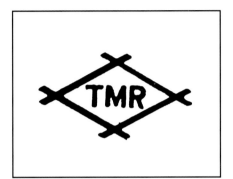

Initials: TMR
Company: **Tomuro** Press Seisakusho
English name: (Tomuro Press Manufacturing)
Trademark derivation: **ToMuRo**

Initials: T.N
Company: **Nomura** Toy Industrial Co., Ltd.
English name: Nomura Toy Industrial Co., Ltd.
Trademark derivation: **N**omura **T**oy (right to left)

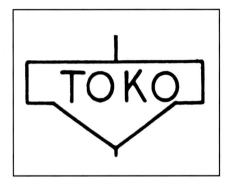

Initials: TOKO
Company: Unknown
English name:
Trademark derivation:

Initials: TOMIYAMA (used 1959-1963)
Company: **Tomiyama** Shoji
English name: Tomiyama Trading Co., Ltd.
Trademark derivation: Eiichiro **Tomiyama**
Company

Initials: TOMY (used 1963-1982)
Company: **Tomy** Company Ltd.
English name: Tomy Company Ltd.
Trademark derivation: Eiichiro **Tomi**yama
Company

Initials: TOMY (used after 1982)
Company: **Tomy** Company Ltd.
English name: Tomy Company Ltd.
Trademark derivation: **Tom**y Company

Initials: Toy Hero
Company: Unknown
English name: Toy Hero
Trademark derivation:

Initials: TOYLAND
Company: **Toyland** Seisakusho
English name: (Toyland Manufacturing)
Trademark derivation: **Toyland**

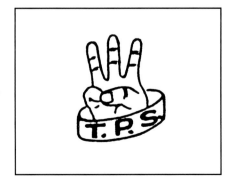

Initials: T.P.S.
Company: **Tokyo Plaything** Shokai
English name: Toplay, Ltd.
Trademark derivation: **T**okyo **P**laything **S**hokai

Initials: TS
Company: **Manmaru** Seisakusho
English name: (Manmaru Manufacturing)
Trademark derivation: Manmaru means Circle
**T**oy, (Mr.) **S**eguchi

Initials: T.S.
Company: **Sato** Gangu Seisakusho
English name: (Sato Toy Manufacturing)
Trademark derivation: Mr. **T**. **S**ato

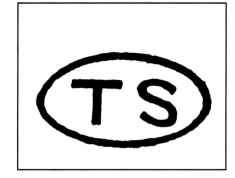

Initials: TS
Company: **Shibuya** Seisakusho
English name: (Shibuya Manufacturing)
Trademark derivation: Mr. **T**oshio **S**hibuya

Initials: T.S.
Company: **Shimazaki** Gangu
English name: (Shimazaki Toy)
Trademark derivation: Mr. **T**. **S**himazaki

Initials: TS
Company: **Taniguchi** Shoten Co., Ltd
English name: Taniguchi Shoten Co., Ltd.
Trademark derivation: (Mr.) **T**aniguchi **S**hoten

Initials: TSS
Company: **Terasawa** Gangu Seisakusho
English name: (Terasawa Toy Manufacturing)
Trademark derivation: **Te**rasawa **S**eisakusho

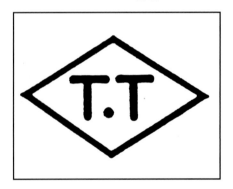

Initials: T.T
Company: **Takatoku** Gangu Co., Ltd
English name: Takatoku Toys Co., Ltd.
Trademark derivation: **T**akatoku **T**oys

Initials: T T
Company: Unknown
English name:
Trademark derivation:

Initials: VIA
Company: **Iwaya** Seisakusho
English name: Iwaya Corporation (a.k.a. **Rock Valley Toys**)
Trademark derivation: VIA derived from Chinese characters. **Iwaya** means **Rock Valley**

Initials: W TOY
Company: **Wakasuto** Boeki
English name: Wakasuto Toy Co. Ltd.
Trademark derivation: **W**akasuto **T**oy

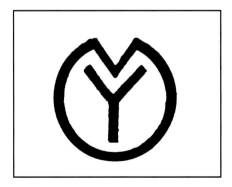

Initials: Y
Company: **Kosei** Gangu Co., Ltd.
English name: Kosei Gangu Co., Ltd.
Trademark derivation: Mr. **Y**oshida

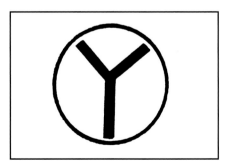

Initials: Y
Company: **Sanko** Seisakusho
English name: (Sanko Manufacturing)
Trademark derivation: Mr. **Y**onezawa

Initials: Y
Company: **Yonezow** Gangu
English name: Yonezow Toys Co., Ltd.
Trademark derivation: Mr. Y. **Y**onezowa

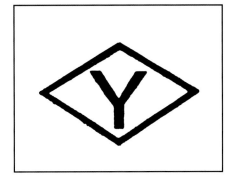

Initials: Y
Company: **Yoshi** Shoji
English name: (Yoshi Trading Company)
Trademark derivation: **Y**oshi

Initials: Y (YO)
Company: **Okabe** Gangu Seisakusho
English name: (Okabe Toy Manufacturing)
Trademark derivation: Mr. **Y**. **O**kabe

Initials: YA + circle
Company: **Maruya** Co., Ltd
English name: Maruya Co., Ltd
Trademark derivation: Maru ya = **circle + Ya**

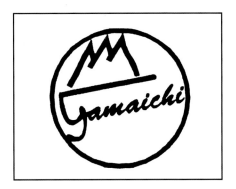

Initials: Yamaichi
Company: **Yamaichi** Shoji
English name: Yamaichi Trading Co.
Trademark derivation: **Yamaichi**

Initials: YF
Company: **Yachio** Seisakusho
English name: (Yachio Manufacturing)
Trademark derivation: (Mr.) **Y**achio **F**ukuda

Initials: YI (IY)
Company: **Yamazaki** Gangu
English name: (Yamazaki Toy)
Trademark derivation: Mr. **I**. **Y**amazaki

Initials: YKS
Company: **Yamamoto** Kinzoku Seisakusho
English name: (Yamamoto Metal Mfg.)
Trademark derivation: **Y**amamoto **K**inzoku **S**eisakusho

Initials: YM
Company: **Morishita** Seisakusho
English name: (Morishita Manufacturing)
Trademark derivation: Mr **Y**. **M**orishita

Initials: YM
Company: **Yanoman** Shoten
English name: (Yanoman Company)
Trademark derivation: **Y**ano**m**an

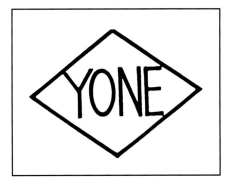

Initials: YONE
Company: **Yoneya** 1964 -
English name: Yoneya Toys Co., Ltd.
Trademark derivation: **Yone**ya

Initials: YY
Company: **Yoshida** Seisakusho
English name: (Yoshida Manufacturing)
Trademark derivation: Mr. **Y**oshio **Y**oshida

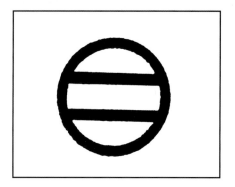

Initials: symbol
Company: **Anzai** Seisakusho
English name: (Anzai Manufacturing)
Trademark derivation: Unknown

Initials: symbol (bull)
Company: **Bull Mark**
English name:
Trademark derivation: Bull

Initials: symbol
Company: **Fuji** Press Kogyo
English name: (Fuji Press Industry)
Trademark derivation: **Mt. Fuji**

Initials: symbol
Company: **Fujii** Seisakusho
English name: (Fujii Manufacturing)
Trademark derivation: Unknown

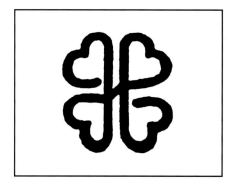

Initials: symbol
Company: **Hayashi** Yushutsu Gangu Kogyo
English name: (Hayashi Export Toy Industry)
Trademark derivation: Unknown

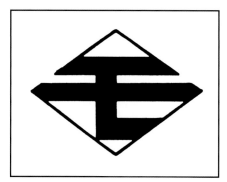

Initials: symbol or HTC
Company: **Hishimo** Sangyo Ltd.
English name: (Hishimo Industrial)
Trademark derivation: **H**ishimo **T**oy **C**ompany

Initials: symbol
Company: **Horikiri** Seisakusho
English name: (Horikiri Manufacturing)
Trademark derivation: Unknown

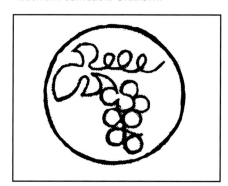

Initials: symbol (Grape Cluster)
Company: **Ichida** Co., Ltd.
English name: Ichida Co., Ltd.
Trademark derivation: Unknown

Initials: symbol (Indian)
Company: **Ichimura** Shoten
English name: Ichimura Co., Ltd
Trademark derivation: Mr. S. Ichimura

Initials: symbol
Company: **Kanasaka** Tekkosho Co., Ltd.
English name: Kanasaka Tekkosho Co., Ltd.
Trademark derivation: Unknown

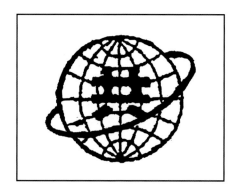

Initials: symbol
Company: **Kyowa** Gangu Seisakusho
English name: (Kyowa Toy Factory)
Trademark derivation: Unknown

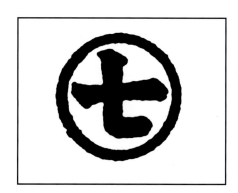

Initials: symbol
Company: **Marushichi** Shokai
English name: Marushichi Co., Ltd
Trademark derivation: Unknown

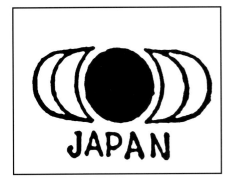

Initials: symbol
Company: **Mitomo** Gangu Co., Ltd.
English name: Mitomo Toy Co., Ltd.
Trademark derivation: Unknown

Initials: symbol
Company: **Mitsuwa** Press Ltd.
English name: Mitsuwa Press Ltd.
Trademark derivation: Unknown

Initials: symbol
Company: **Momoya** Shoten
English name: (Momoya Co.)
Trademark derivation: Unknown

Initials: symbol - Chinese
Company: **Nakasho** Kojyo Co., Ltd.
English name: Nakasho Kojyo Co., Ltd.
Trademark derivation: Mr. **Naka**mura **Sho**toro

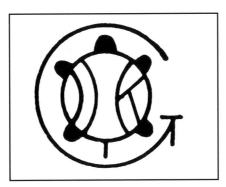

Initials: symbol
Company: **Nikko** Gangu Kogyo
English name: (Nikko Toy Industry)
Trademark derivation: Unknown

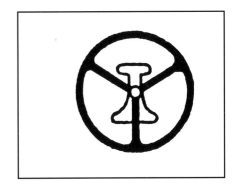

Initials: symbol
Company: **Nishikata** Gangu Kogyo
English name: (Nishikata Toy Industry)
Trademark derivation: Unknown

Initials: symbol (Turtle)
Company: **Nomura** Toy Industrial Co., Ltd.
English name: Nomura Toy Industrial Co., Ltd.
Trademark derivation: Unknown

Initials: symbol
Company: **Ohara** Seisakusho
English name: (Ohara Manufacturing)
Trademark derivation: Unknown

Initials: symbol
Company: **Okuma** Seisakusho
English name: (Okuma Manufacturing)
Trademark derivation: Unknown

Initials: symbol
Company: **Sanko** Seiki Co., Ltd
English name: Sanko Seiki Co., Ltd
Trademark derivation: Unknown

Initials: symbol
Company: **Sanshin** Toys Co., Ltd.
English name: Sanshin Toys Co., Ltd.
Trademark derivation: Unknown

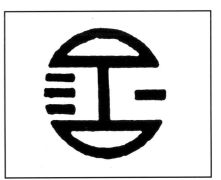

Initials: symbol
Company: **Shinkosha** Co., Ltd
English name: Shinkosha Co., Ltd
Trademark derivation: Unknown

Initials: symbol
Company: **Shinsei** Kogyo
English name: (Shinsei Industry)
Trademark derivation: Unknown

Initials: symbol
Company: **Suga** Co., Ltd.
English name: Suga Co., Ltd.
Trademark derivation: Unknown

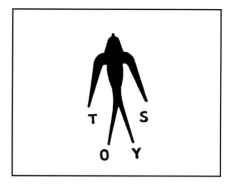

Initials: symbol + TOYS
Company: **Swallow** Toys Co., Ltd.
English name: Swallow Toys Co., Ltd.
Trademark derivation: **Swallow + TOYS**

Initials: symbol
Company: **Taiseiya** Co., Ltd.
English name: Taiseiya Co., Ltd.
Trademark derivation: Unknown

Initials: symbol
Company: **Taiyo** Kogyo
English name: (Taiyo Industry)
Trademark derivation: Unknown

Initials: symbol
Company: Unknown
English name:
Trademark derivation: Unknown

Initials: symbol (Rabbit)
Company: **Usagiya**
English name: Usagiya Toys Co., Ltd.
Trademark derivation: **Usagi means rabbit**

Initials: symbol
Company: **Yoshino** Kogyo
English name: Yoshino Industrial Co., Ltd.
Trademark derivation: Unknown

# SCARCITY, CONDITION, AND VALUES

## SCARCITY

Value is directly related to scarcity, with scarcity being determined by the *rarity* of the toy, *demand* for the toy, and finally, the *condition* of the toy. Each toy has been given a scarcity rating on a scale of 1 to 5 with 5 being the most scarce. I chose to use scarcity as a measure because it takes into account not only how many were produced but how much of a demand exists for the toy among collectors. A rare toy with low production quantities can exist without much demand, just as a more plentiful toy can be in high demand and relatively more scarce.

## RARITY

Rarity is influenced primarily by how many toys were produced, how many survived, and in what part of the world they were marketed. To the manufacturer, the success of a toy was measured by sales. Poor selling toys were a financial disappointment to the manufacturer, but because of low supply and high demand, they result in higher prices for the collector of today. Based on the recollection of Mr. Udagawa and other long time Toplay employees, an estimate of the production quantities for most toys has been included in the individual descriptions found in this book. This will give the reader a starting point in understanding the relative availability of the toy. However, it does not take into account where these toys were sold.

Some toys were only produced as samples that were taken by the trading companies to show importers what was available. Importers may have taken a sample but never ordered a supply, thus the toy never went into higher volume production. Some toys were immensely popular, resulting in large quantity production. The "Roller Skating Clown" is a good example of this success. Other toys were not well received, resulting in very minimal production. The "Animal Barbershop" did not sell particularly well and was a "flop" in the eyes of Mr. Udagawa, even though it is a well designed toy that is very popular today. Although accurate production records do not exist, having some idea of the approximate number produced is a piece of information not often available to collectors. Please keep in mind these quantities are approximate. The dates shown for the introduction of the toy were also based on recollection, but assisted by catalogs and patent application records that were available.

Individuals who had these toys as children, and their parents, did not realize they would become so collectible and valuable someday. Accordingly, toys were given away, thrown away, or generally abandoned due to lack of interest. It is probably reasonable to assume that less than 10-15% of the earlier toys have survived over time. If you apply that survival rate to the amount produced, you start to have a good sense of how rare any given toy may be.

Even though we live in a very global and mobile society, the location where the toys were marketed impacts where they are today. The result is that toys made primarily for the North American market are much harder to find in other parts of the world, causing the prices realized for those toys in other areas to be higher than in North America. Conversely, toys made for the Japanese market are not often found in North America. This makes these toys rarer in North America with the prices higher accordingly. Most of the Japanese toys of this period were produced for export as opposed to the Japanese market.

With regard to T.P.S., probably 60% of the production ended up in North America, 25-30% in Europe, and the balance to the rest of the world. However, it is important to note that this varied by individual toy, as all toys were not sold to the same buyers in the same countries. Toy boxes were printed only in the English or Japanese language.

## DEMAND

The next major influence on values is demand. Some toys are just more popular than others, based on the interests of collectors. Though not limited to character toys, this is probably the best example of popularity; toys representing popular and known characters are often more sought after by collectors. Saying it in a different way, collectors in these categories specifically seek out representations of their favorite characters, which then increases the demand for that particular toy. Examples of T.P.S. character toys would include the globally popular Popeye or Disney characters. More popular in Japan, but still globally recognized, would be Tetsuwan Atom or Astroboy characters, Tetsujin 28 or Gigantor, Keroyon, Uncle Geba-Geba, and Grendizer.

Other toys are more popular because of their subject material. Clowns, animals, motorcycles, planes, cars, trains, robots, and space toys are all examples of subject collecting. Logic follows that toys sought by more collectors end up in more collections, making those toys harder to find.

# CONDITION & GRADING

Toys have survived in all stages of condition, from junk parts to still near mint and in the original box. While the grading scale of C-1 to C-10 is widely utilized, there is certainly little consistency on how individuals grade their toys against this scale. If C-10 is truly mint (just as it came from the factory) and C-1 is parts, most collectors will search out toys in the C-6 to C-10 range. There is minimal interest among collectors for toys below C-6 (*Fine*) condition. Accordingly, value estimates have been shown only for C-6, C-8, and C-10 toys, all without boxes.

C-6: *Fine.* Complete, with no missing or broken parts. Nice condition with some evidences of aging and wear, but not played with hard.

C-7: *Very Fine.* Very minimal scratching and wear, but still bright.

C-8: *Excellent.* Very light general wear and appears close to new.

C-9: *Near Mint.* Looks like new, but upon close examination is not truly mint.

C-10: *Mint.* As it originally came from the factory with no defects, however factory touch-up is acceptable.

Finding toys that are in factory mint condition is very difficult and these toys are uncommon! The best chance for this happening is when someone finds old unused store stock. With T.P.S. toys, often the clothing was the most vulnerable to aging. The satin-like material used for the pants was particularly subject to color fading with age, as well as to staining from the clockwork mechanism. The felt or velvet material often used for other clothing was also subject to the same likelihood of fading. While the rest of the toy may appear mint, this fading keeps the toy from being considered a C-10.

Boxes often suffered a similar fate. Because these toy boxes were not very strong, they were very susceptible to tearing and bending with handling. Toy protrusions such as keys would often poke a hole in the box. Moisture would cause the box staples to rust and children would draw on the boxes. Still, the box art on many of the toys is very colorful and pictorial, making the boxes a significant item to the collector and very desirable in their own right. Accordingly, boxes should be graded separately to accurately describe the condition of both the toy and box and to better determine the approximate value of the toy and its box together. The C-1 to C-10 scale is appropriate for box grading. Sometimes a C-10 box with C-10 toy can almost double the value of the toy, however a C-6 box may only add 20% to the value of the toy.

# VALUES

Current values for each toy are shown with the individual toy identification. Establishing values ranges is by no means an exact science, due to the many variables and subjectivity involved. In the end, the market will determine the price realized at a sale, so the guide should be viewed as just that — a guide! The values published in this book represent a compilation of known sales at auction houses, Internet auctions, and the sales at antique and specialized toy shows. In most cases they correspond to rarity of the toy and the market demand. The prices represent primarily North American values and should be considered in light of my comments above regarding interest and prices in other major geographic markets such as Japan and Europe.

It is not uncommon to find boxed tin windup toys from the 1950s and 1960s with the original price tag still intact. A price of 79¢ or 89¢ is typical. But alas, this is the new millennium and the year 2000 brings us selling prices of over 600 times that amount! In fact, it is hard to find collector's condition (C-6 to C-10) T.P.S. toys for under $100.

# ALPHABETICAL LISTING OF TOYS

The pages that follow are organized in alphabetical order by actual toy name. To assist in identifying these names, an index of toys by subject category with thumbnail photographs is included for your reference on pages 208-224. Each listing in this section contains photographs of the toy plus its box where possible, followed by:

• Description of the toy, including mechanism and composition
• Toy marks and locations
• Dimensional data (Measurements were made in metric units since this was the original unit used in the design and manufacturing of the toy. Metric units were converted to inches and rounded off to the nearest 0.25 inch. Both are shown for convenience.)
• Approximate introduction date
• Approximate production quantity
• Scarcity rating
• Current values for C-6, C-8, and C-10 toys
• Box text and markings
• Description of the toy operation, including action illustration drawings or photos when available
• Selected outline and patent drawings

## ACROBAT TEAM PORSCHE

Box text: "BATTERY OPERATED ACROBAT TEAM PORSCHE." Box marks: T.P.S.

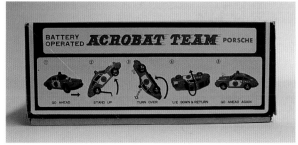

**Action:** Car goes forward. Front end lifts up via lever on bottom of car and turns car over backwards. The weight in the car then causes it to roll over on its right side and upright itself going in the opposite direction.

Battery operated TURNOVER 911-S PORSCHE, POLICE or FIRE DEPARTMENT CHIEF. Tin with plastic wheels. Known variations: Police, Fire Chief, or Racing lithography.
Toy Marks: T.P.S. (right side car door).
Size: L: 10in H: 4in W: 4in (L: 25cm H: 10cm W: 10cm)
Introduced: c.1977. Est. quantity: 30,000 pieces each. Scarcity rating: 2.5.
*Fine to Excellent (C6 to C8): $75-100, Mint (C10): $125.*

Fire Department Chief Porsche.

# ANIMAL BARBER SHOP

Windup MOUSE BARBER SHAVING A RABBIT with fixed key. Tin with rubber ears.
Size: H: 5in L: 4.75in W: 2.75in (H: 12.5cm L: 12cm W: 7cm)
Introduced: c.1955. Est. quantity: 12,000 pieces. Scarcity rating: 4.
*Fine to Excellent (C6 to C8): $300-$400, Mint (C10): $525.*

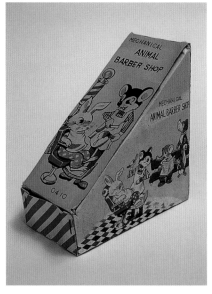 

Box text: "MECHANICAL ANIMAL BARBER SHOP - 0410." Box marks: T.P.S.
**Action:** The barber moves his head back and forth with a razor moving in his right hand and a shave brush in his left hand.

Toy Marks: T.P.S. (right rear leg), Hishim◌ (left rear leg).

# ANIMAL'S PLAYLAND

Box text: "MECHANICAL ANIMAL'S PLAYLAND," "RUNS LONG 2 1/2 MINUTES DURABLE." Box marks: T.P.S.

Windup RABBIT SWINGING FROM TREE WITH ANIMAL HOUSE SIGN. BEAR & MONKEY ON SEE-SAW, with fixed key. Tin.
Toy Marks: T.P.S. (back side of tree house).
Size: L: 9.75in H: 5in D: 3.25in (L: 24.5cm H: 13cm W: 8.5cm)
Introduced: c.1967. Est. quantity: 12,000 pieces. Scarcity rating: 4.
*Fine to Excellent (C6 to C8): $225-$300, Mint (C10): $375.*

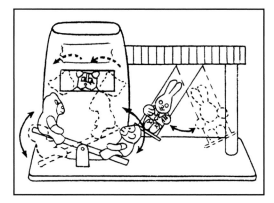

**Action:** Rabbit swings, see-saw moves, and three different animals appear through the window in the tree house.

# APOLLO SPACE PATROL WITH SATELLITE SHIP

Box text: "BATTERY OPER-ATED APOLLO SPACE PATROL WITH SATELLITE SHIP," "MYSTERY ACTION * THE LIGHT FORWARD RED GREEN AND YELLOW LIKE NEON SIGNS * THE SATEL-LITE SHIP REVOLVES ON THE TOP * REALISTIC ENGINE SOUND * 95100." Box marks: T.P.S.

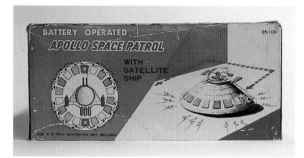

Battery operated FLYING SAUCER WITH ROTATING MANNED SATELLITE SHIP AND WING ROCKETS ON TOP. Plastic saucer with tin and plastic satellite ship.
Toy Marks: T.P.S. (left wing of satellite ship).
Size: Diameter: 8.75in H: 5.5in (Diameter: 22cm H: 14cm)
Introduced: c.1970. Est. quantity: 12,000 pieces. Scarcity rating: 4.
*Fine to Excellent (C6 to C8): $200-$250, Mint (C10): $350.*

**Action:** Bump and go saucer with counter rotating sets of sequentially flashing lights. Satellite ship with gun and rockets spins on top.

# BALL PLAYING GIRAFFE

Windup GIRAFFE SITTING AND BOUNCING BALL, with separate key. Tin with tin ears and rubber antlers. Known variations: LINEMAR version with LINEMAR mark on toy and box.
Toy Marks: Made in Japan (base, in front of giraffe).
Size: H: 8.5in L: 4.25in W: 2.5in (H: 22cm L: 11cm W: 6.5cm)
Introduced: c.1958. Est. quantity: 50,000 pieces. Scarcity rating: 3.
*Fine to Excellent (C6 to C8): $175-$225, Mint (C10): $300.*

Box text: "MECHANICAL BALL PLAYING GIRAFFE," "After winding - lift the ball and then drop into Giraffe's hands." Box marks: T.P.S.

**Action:** Sitting giraffe bounces ball on a wire between its face and front feet, which move up and down.

# BANJOIST

Windup STRUMMING COWBOY BANJO PLAYER with fixed key. Tin.
Toy Marks: LINEMAR (back, below key).
Size: H: 5.25in W: 2.25in D: 1.5in (H: 13.5cm W: 6cm D: 4cm)
Introduced: c.1956. Est. quantity: 10,000 pieces. Scarcity rating: 4.
*Fine to Excellent (C6 to C8): $175-$225, Mint (C10): $300.*

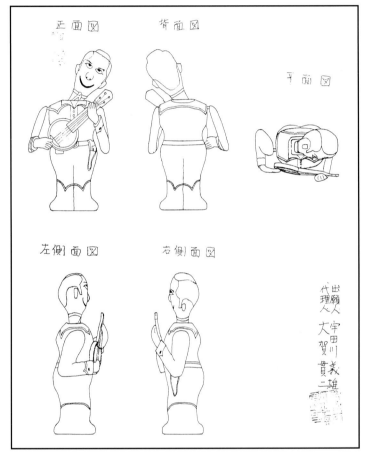

正面図　　　背面図　　　平面図

左側面図　　　右側面図

出願人
代理人
宇田川
大賀　貫義
二雄

"Banjoist" Drawing. Japanese Patent # 128040 granted 1957.
*Courtesy of Toplay (T.P.S.) Ltd.*

ONE OF THE MANY LINEMAR TOYS DO YOU HAVE ALL OF THEM ?

**Action:** Man holds banjo in left hand while right arm moves back and forth across the banjo in realistic strumming action. Body vibrates and moves while head moves left and right.

Box text: "MECHANICAL BANJOIST." Box marks: LINEMAR.

# BARREL ROLL RACE CAR SET WITH SEESAW

Windup END TO END PIVOTING RAMP RACE TRACK WITH TWO SIDED ROLLOVER RACE CAR and separate key. Plastic with tin race car with rubber wheels.
Toy Marks: T.P.S. (rear of race car).
Size: L: 43in H: 4.75in W: 3.25in (L: 109cm H: 12cm W: 8cm)
Introduced: c.1970. Est. quantity: 24,000 pieces. Scarcity rating: 4.
*Fine to Excellent (C6 to C8): $50-$60, Mint (C10): $75.*

Box text: "MECHANICAL BARREL ROLL RACE CAR SET WITH SEESAW," "TURNS UPSIDE DOWN AND STARTS AGAIN * BUMPS AGAINST TIRES AND PASSES THROUGH GATE * SEESAWS." Box marks: T.P.S.

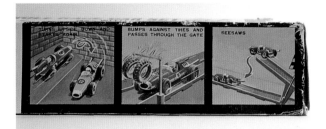

**Action:** Windup 9.5cm (3.75in) race car climbs ramp which pivots, sending car to the other side. Shape of track causes car to barrel roll 180° allowing the car to reverse direction to repeat action without having to reverse wheels.

# BEAR GOLFER

Windup BEAR HITTING GOLF BALL INTO NET AND HOLE, with fixed key. Includes two white balls. Tin with netting and rubber ears.
Size: H: 4.25in L: 7.75in W: 2.75in (H: 11cm L: 19.5cm W: 7cm)
Introduced: c.1959. Est. quantity: 100,000 pieces. Scarcity rating: 2.
*Fine to Excellent (C6 to C8): $175-$225, Mint (C10): $275.*

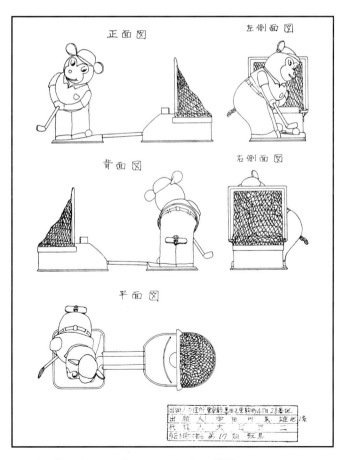

Drawing from Japanese Patent application, 1959.
*Courtesy of Toplay (T.P.S.) Ltd.*

Box text: "MECHANICAL BEAR GOLFER,"
"SHOOTS A HOLE-IN-ONE EVERY TIME * PAT.
PEND." Box marks: T.P.S., Hirata.

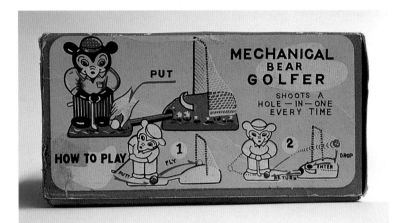

Toy Marks: T.P.S. (right rear pant leg), Hishimo (left rear pant leg).
Also seen with Hirata logo.

**Action:** Bear swings golf club hitting ball into net which
drops into hole and returns. Has Start/Stop switch.

## BEAR PLAYING BALL

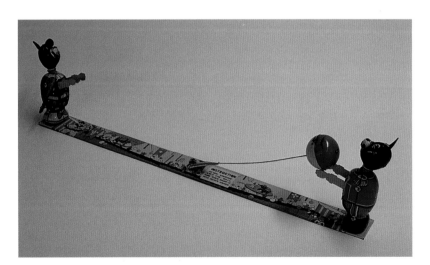

Box text: "MECHANICAL BEAR PLAYING BALL."
Box marks: T.P.S.

TWO WIND-UP BEARS THROWING BALL BACK AND FORTH with fixed
key on each bear. Tin with rubber ears on bears.
Toy Marks: T.P.S. (right side of base).
Size: L: 19in H: 4.25in W: 1.5in (L: 48cm H: 11cm W: 4cm)
Introduced: c.1957. Est. quantity: 20,000 pieces. Scarcity rating: 3.5.
*Fine to Excellent (C6 to C8): $250-$300, Mint (C10): $375.*

**Action:** Winding each bear causes arms to bounce ball on
wire, back and forth between bears. Base folds for storage

# BIG LEAGUE HOCKEY PLAYER

Windup SKATING HOCKEY PLAYER WITH DETACHABLE HOCKEY STICK AND PUCK and fixed key. Tin with rubber wheels. Toy Marks: Made in Japan (right hip). Size: H: 6.25in W: 2.75in L: 7.75in (H: 16cm W: 7cm L: 20cm) Introduced: c.1957. Est. quantity: 12,000 pieces. Scarcity rating: 4. *Fine to Excellent (C6 to C8): $275-$375, Mint (C10): $500.*

Box text: "MECHANICAL BIG LEAGUE HOCKEY PLAYER * NO. 6252." Box marks: T.P.S., AHI.

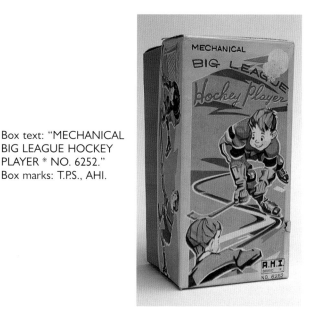

**Action:** Hockey player skates forward on wheels, turns, and swings hockey stick. Stick is removable for storage.

# BIG STUNT CAR

Box text: "BATTERY OPERATED BIG STUNT CAR," "GO AHEAD * INCLINE * TURN OVER * GO AHEAD AGAIN." Box marks: T.P.S.

Battery operated ROLL OVER BLACK 1970 MACH I MUSTANG WITH INTE-RIOR AND GOLD GRILL AND BUMPERS. Tin with 1 rubber and 3 plastic tires. Toy Marks: T.P.S. (rear window deck). Size: L: 10.75in H: 3.25in W: 4.25in (L: 27cm H: 8cm W: 11cm) Introduced: c.1975. Est. quantity: 30,000 pieces. Scarcity rating: 3. *Fine to Excellent (C6 to C8): $100-$125, Mint (C10): $150.*

**Action:** Car goes forward, raises up on two side wheels via cam and lever, then barrel rolls 360° before going forward again. Front wheels can be positioned for car direction.

# BOBO THE JUGGLING CLOWN

Windup CLOWN JUGGLING BALL BETWEEN ARMS, with fixed key. Tin.
Toy Marks: T.P.S. (on base between legs).
Size: H: 5in W: 6.75in D: 1.5in (H: 13cm W: 17cm D: 4cm)
Introduced: c.1956. Est. quantity: 12,000 pieces. Scarcity rating: 4.
*Fine to Excellent (C6 to C8): $425-$550, Mint (C10): $700.*

Box text: "BOBO THE MECHANICAL JUGGLING CLOWN"
Box marks: T.P.S. *Brynne and Scott Shaw Collection.*

**Action:** Arms are spring loaded after winding. Dropping the
wire with attached ball against one arm starts the ball bouncing
back and forth between the clown's outstretched arms in
juggling fashion. *Brynne and Scott Shaw Collection.*

# BOUNCING BALL DOLLY

Windup GIRL KNEELING AND BOUNCING BALL WITH RIGHT
HAND, with fixed key. Tin with vinyl head.
Toy Marks: T.P.S. (back, left side above shoe).
Size: H: 5.25in W: 2.75in D: 4cm (H: 13.5cm W: 7cm D: 10cm)
Introduced: c.1961. Est. quantity: 20,000 pieces. Scarcity rating: 3.
*Fine to Excellent (C6 to C8): $125-$150, Mint (C10): $200.*

Box text: "MECHANICAL BOUNCING BALL DOLLY."
Box marks: T.P.S.

**Action:** Kneeling girl scoots across the surface on
wheels, while bouncing ball on wire between surface and
her moving right hand.

# BRAVE HOVERCRAFT

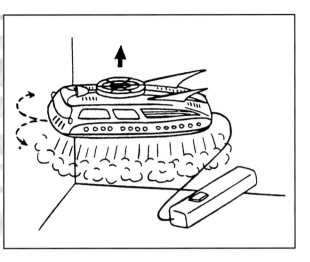

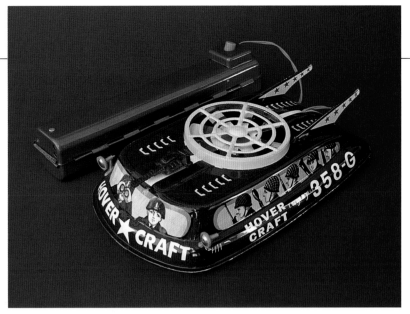

Battery operated ARMY GREEN FLOATING HOVERCRAFT 358G WITH REMOTE BATTERY CONTROL. Aluminum with plastic gun and tin battery holder.
Toy Marks: T.P.S. (right rear of Hovercraft).
Size: L: 7.75in W: 4.75in H: 2.25in (L: 20cm W: 12cm H: 5.5cm)
Introduced: c.1968. Est. quantity: 10,000 pieces. Scarcity rating: 4.5.
*Fine to Excellent (C6 to C8): $300-$400, Mint (C10): $500.*

Box text: "BATTERY OPERATED BRAVE HOVERCRAFT," "NO WHEELS * DRIVING FORCE BY STRONG EXHAUSTION * RISES, HOVERS, DARTS ABOVE SURFACE." Box marks: T.P.S.

**Action:** Hovercraft rises, hovers and darts above the surface as rotating blades inside create air movement to lift the lightweight aluminum body off the ground.

# BUNNY FAMILY PARADE

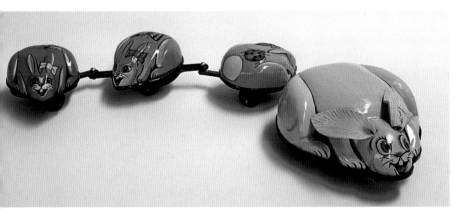

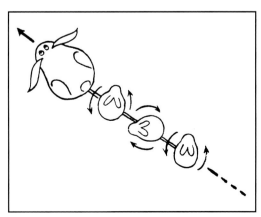

Friction MAMA BUNNY WITH THREE BABY BUNNIES FOLLOWING IN SINGLE LINE. Tin with rubber ears and wheels.
Toy Marks: T.P.S. (rear of Mama bunny near tail).
Size: L: 12.5in W: 3.25in H: 1.75in (L: 32cm W: 8cm H: 4.5cm)
Introduced: c.1959. Est. quantity: 20,000 pieces. Scarcity rating: 3.5.
*Fine to Excellent (C6 to C8): $110-$125, Mint (C10): $150.*

Box text: "FRICTION BUNNY FAMILY PARADE," "SEE THE BABY BUNNIES CAPER WHILE FOLLOWING MAMA BUNNY." Box marks: T.P.S., Hikari.

**Action:** Friction movement with attached baby bunnies following mama and spinning around individually.

## BUSY BUG

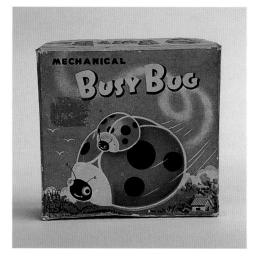

Windup ZIGZAG LADY BUG WITH BABY PICTURED ON ITS BACK and fixed key. Tin with rubber wheels and spring antennae.
Toy Marks: T.P.S. (back of bug).
Size: L: 4in H: 2in W: 3.25in (L: 10cm H: 5cm W: 8cm)
Introduced: c.1959. Est. quantity: 10,000 pieces. Scarcity rating: 4.
*Fine to Excellent (C6 to C8): $50-$70, Mint (C10): $95.*

Box text: "MECHANICAL BUSY BUG." Box marks: T.P.S.

**Action:** Lady bug scurries around in a busy manner, alternating directions via oscillating front drive wheel.

## BUSY CHOO CHOO

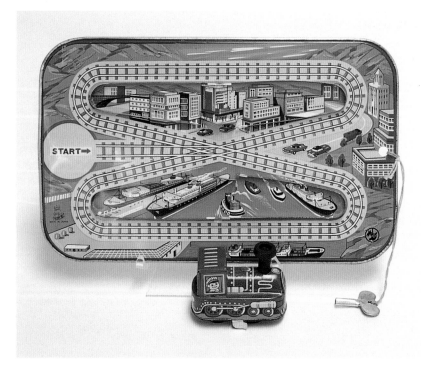

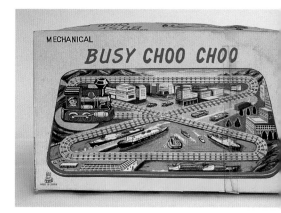

Windup 7cm (2.75in) ENGINE ON PLATFORM BASE WITH CITY AND PORT SCENES, with separate key. Tin with rubber stack on loco.
Toy Marks: T.P.S. (lower left side of base), Hirata (lower right corner of base).
Size: 9.25in x 5.5in platform (23.5cm x 14cm platform)
Introduced: c.1967. Est. quantity: 50,000 pieces. Scarcity rating: 2.
*Fine to Excellent (C6 to C8): $75-$95, Mint (C10): $125.*

Box text: "MECHANICAL BUSY CHOO CHOO." Box marks: T.P.S., Franconia.

Packaging variation: Bag packaging marked "T.P.S. P-216A," "LOTS OF SPEEDY ACTION ON THE RAIL." This version has base mounted keywind with instructions.

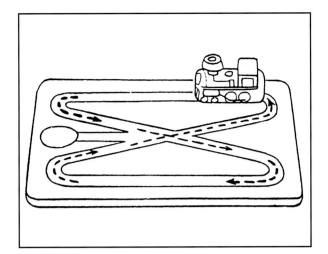

**Action:** Windup engine goes around railroad platform hooked on guide rail in figure eight route.

# BUSY EMERGENCY CAR SERIES

Battery operated STEERABLE COMIC POLICE CAR and FIRE CAR. Tin with plastic figure and wheels. Variations: Black and white Police or red Fire car.
Toy Marks: T.P.S. (rear, above tail lights).
Size: L: 7.5in W: 4.25in H: 6.25in (L: 19cm W: 11cm H: 16cm)
Introduced: c.1970. Est. quantity: 1,200 pieces. Scarcity rating: 5.
*Fine to Excellent (C6 to C8): $150-$200, Mint (C10): $275.*

Box text: "BATTERY OPERATED BUSY EMERGENCY CAR SERIES," "PLACE BATTERY IN BOTTOM OF CHASSIS. TURN ON SWITCH AND SET WHEELS IN DIRECTION DESIRED. PLACE ON FLOOR AND LET GO." Box marks: F.E.White.

**Action:** Set wheels in direction desired. Irregular axle causes driver to slide around on seat. Hands move on steering wheel and mouth opens and closes.

## BUSY F.D. LADDER TRUCK

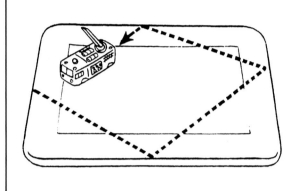

Windup 7cm (2.75in) FIRE ENGINE ON BURNING BUILDING SCENE PLATFORM BASE MARKED F.D. LADDER TRUCK, with fixed key. Tin.
Toy Marks: T.P.S. (base, lower right corner).
Size: 9.25in x 6.25in platform (23.5cm x 15.5cm platform)
Introduced: c.1965. Est. quantity: 50,000 pieces. Scarcity rating: 2.5.
*Fine to Excellent (C6 to C8): $85-$110, Mint (C10): $140.*

Box text: "MECHANICAL BUSY F.D. LADDER TRUCK - MYSTERY NON STOP ACTION." Box marks: T.P.S.

**Action:** Windup fire engine with mystery bump and go action on platform. Fire engine stays within the confines of flat rail on base via knob protruding from underside.

## BUSY MOUSE

Box text: "MECHANICAL BUSY MOUSE * MYSTERY NON-STOP ACTION." Box marks: T.P.S.

Windup 9cm (3.5in) MOUSE ON PLATFORM BASE WITH TOM AND JERRY LIKE KITCHEN SCENES, with fixed key. Tin with plastic ears and rubber tail on mouse.
Toy Marks: T.P.S. (base, lower left corner).
Size: 9.25in x 6.25in platform (23.5cm x 15.5cm platform)
Introduced: c.1965. Est. quantity: 30,000 pieces. Scarcity rating: 3.
*Fine to Excellent (C6 to C8): $85-$110, Mint (C10): $135.*

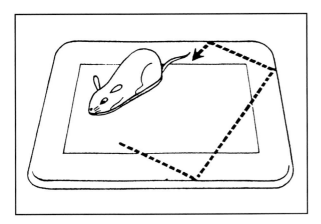

**Action:** Mouse moves within flat rail around platform base with bump and go action initiated by knob protruding from bottom of mouse.

Packaging variation: Bag packaging.

# CALYPSO JOE THE DRUMMER

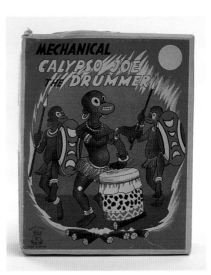

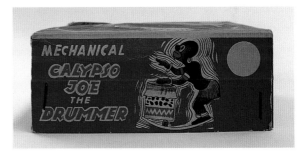

Windup NATIVE IN GRASS SKIRT BEATING DRUM AND ROCKING BACK AND FORTH, with fixed key. Tin with rubber hands, fringed waist, cuff, and drum. Known variations: T.P.S. and LINEMAR version.
Toy Marks: T.P.S. (Joe's right side above waist) or LINEMAR (Joe's right side above waist).
Size: L: 4.5in H: 5.5in W: 2.25in (L: 11.5cm H: 14cm W: 5.5cm)
Introduced: c.1960. Est. quantity: 30,000 pieces. Scarcity rating: 3.5.
*Fine to Excellent (C6 to C8): $300-$365, Mint (C10): $450.*

Box text: "MECHANICAL CALYPSO JOE." Box marks: T.P.S. or LINEMAR.

**Action:** Joe bends over and beats the drum with his moving arms, then stands up straight before repeating the action. This causes the rocker to move back and forth.

Box side panel.

# CANDY LOVING CANINE

Toy seen with and without H (Hirata) mark.

Windup DOG THROWING CANDY INTO MOUTH. COMES WITH 3 CANDY BALLS and fixed key. Tin with rubber ears and plastic candy balls.
Toy Marks: T.P.S. (back left side), Hirata (back left side).
Size: H: 5.5in W: 2in D: 3.25in (H: 14cm W: 5cm D: 8cm)
Introduced: c.1959. Est. quantity: 50,000 pieces. Scarcity rating: 2.5.
*Fine to Excellent (C6 to C8): $125-$150, Mint (C10): $190.*

Box text: "MECHANICAL CANDY LOVING CANINE," "ALWAYS HUNGRY NEVER SATISFIED."
Box marks: T.P.S., Hirata.

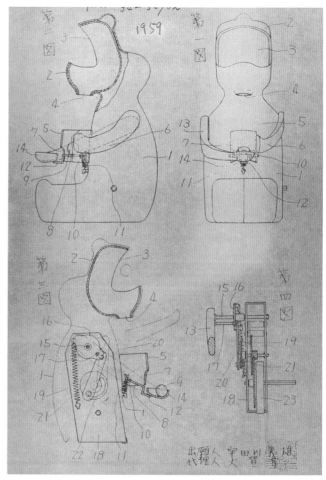

1959 Japanese Patent drawing. *Courtesy of Toplay (T.P.S.) Ltd.*

**Action:** Sitting dog takes small candy-like ball from can of candy and throws into mouth. Candy drops from hole in mouth back into can. Includes start/stop switch.

# CAT & MOUSE IN SHOE WITH VOICE

Windup CAT AND MOUSE POPPING IN AND OUT OF SHOE, with fixed key. Tin.
Size: L: 6.5in (L: 16.5cm)
**Action:** Shoe moves around in erratic action while cat and mouse raise their heads alternately as if hiding from each other. Squeak box simulates cat mewing during action.
Introduced: c.1965. Est. quantity: 6,000 pieces. Scarcity rating: 5.
*Fine to Excellent (C6 to C8): $175-$250, Mint (C10): $350.*

# CATERPILLAR BULLDOZER

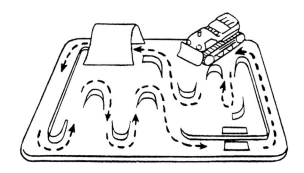

Windup 6cm (2.25in) BULLDOZER ON PLATFORM BASE WITH CONSTRUCTION SCENES AND TUNNEL, with base mounted key. Tin. Known variations: Marketed also as "Zigzag Bulldozer."
Toy Marks: T.P.S. (lower left corner of base), Hirata (lower right corner of base).
Size: 15.25in x 6.75in platform (38.5cm x 17cm platform)
Introduced: c.1967. Est. quantity: 40,000 pieces. Scarcity rating: 3.
*Fine to Excellent (C6 to C8): $100-$120, Mint (C10): $150.*

Box text: "MECHANICAL CATERPILLAR BULLDOZER." Box marks: T.P.S., Hirata.

**Action:** Windup bulldozer goes around platform and through tunnel hooked on zigzag rail.

# CHAMP ON ICE

Windup THREE BEARS SKATING TOGETHER IN SINGLE FILE, with fixed key. Tin.
Toy Marks: T.P.S. (right side of leg, lead skater).
Size: L: 8.75in H: 5in W: 2in (L: 22cm H: 13cm W: 5cm)
Introduced: c.1960. Est. quantity: 12,000 pieces. Scarcity rating: 4.
*Fine to Excellent (C6 to C8): $550-$725, Mint (C10): $900.*

Box text: "CHAMP ON ICE," "BEAR SKATER TRIO * WIND-UP CLOCK MECHA-
NISM." Box marks: T.P.S. *Brynne and Scott Shaw Collection.*

**Action:** Three bears skating in line with complex mechanism. Front bear swings arms
as other bears kick their left legs in skating motion. Wheels on first and last skater.
*Brynne and Scott Shaw Collection.*

Box variation from Greece. Toy inside was
identical to or copy of "Champ on Ice." *Photo
courtesy of New England Auction Gallery.*

# CHAMPION AUTO RACE TOURNAMENT

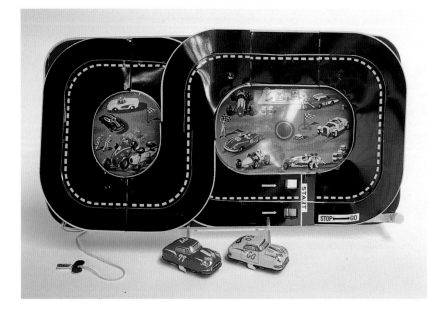

TWO 6.5cm (2.5in) WINDUP CARS RACE ON OVER/
UNDER FIGURE EIGHT RACE TRACK ON PLAT-
FORM BASE WITH RACING SCENES. Includes
separate key. Tin with rubber front wheels on cars.
Toy Marks: T.P.S. (lower left side of base), Hirata (lower
left side of base).
Size: L: 16.75in W: 8.75in H: 2in (L: 42.5cm W: 22cm
H: 5cm)
Introduced: c.1965. Est. quantity: 10,000 pieces.
Scarcity rating: 4.
*Fine to Excellent (C6 to C8): $135-$160, Mint (C10): $200.*

Box text: "MECHANICAL CHAMPION AUTO RACE TOURNA-MENT." Box marks: T.P.S.

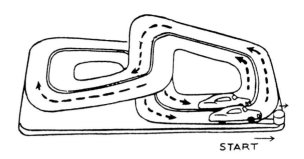

**Action:** Cars are placed at starting line where gravity, gear stopping mechanisms drop in slots on track. Pulling the "go" lever raises the mechanism and starts the cars on their race. Tabs under cars are hooked on to guide rails to maintain proper lanes.

# CHAMPION STUNT CAR

Battery operated ROLL OVER SILVER 1970 MACH I MUSTANG WITH STUNT RACING LITHO AND BLUE TINTED WINDOWS. Tin with I rubber tire and 3 plastic tires.
Toy Marks: T.P.S. (top of left rear fender).
Size: L: 10.75in H: 3.25in W: 4.25in (L: 27cm H: 8cm W: 11cm)
Introduced: c.1979. Est. quantity: 12,000 pieces. Scarcity rating: 3.5.
*Fine to Excellent (C6 to C8): $100-$125, Mint (C10): $175.*

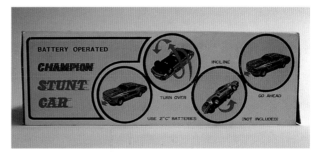

Box text: "BATTERY OPERATED CHAMPION STUNT CAR," "GO AHEAD * INCLINE * TURN OVER." Box marks: T.P.S.

**Action:** Car goes forward in circular pattern and raises up on two side wheels via cam and lever mechanism. Car then barrel rolls 360° and goes forward again. Fixed front wheels.

Champion Stunt Car action mechanism.

# CIRCUS ACROBATIC SEAL AND BALL

Windup SEAL WITH MOVING REAR FLIPPERS, BOUNCING BALL ON NOSE, with fixed key. Version 1 (left): Tin with satin like collar and pants, and plastic ball. Version 2 (right): Same seal with different litho printing and felt coat, which changes appearance significantly. Also seen with tin multicolored ball.
Toy Marks: T.P.S. (right side of chest).
Size: H: 5.5in W: 3.25in D: 4in (H: 14cm W: 8cm D: 10cm)
Introduced: c.1957 1959. Est. quantity: 6000 pieces (version 1), 12,000 pieces (version 2). Scarcity rating: 4.
*Fine to Excellent (C6 to C8): $95-$120, Mint (C10): $150.* (Version 1, add 15-20%).

Box text: "CIRCUS-ACROBATIC SEAL and BALL MECHANICAL," "CRAGSTAN NO.566." Box marks: T.P.S., Cragstan.

**Action:** Seal walks forward and bounces ball on its nose. Ball is connected by rod. Rear half of body, with flippers, moves to create walking motion.

# CIRCUS BUGLER

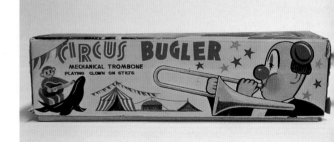

Windup SWAYING, TROMBONE PLAYING CLOWN ON STILTS WITH STAR EYED FACE and fixed key. Tin with cloth pants, felt bow tie and jacket.
Toy Marks: None.
Size: H: 9.5in W: 2.25in D: 3.25in (H: 24cm W: 5.5cm D: 8cm)
Introduced: c.1955. Est. quantity: 10,000 pieces. Scarcity rating: 4.
*Fine to Excellent (C6 to C8): $325-$425, Mint (C10): $550.*

Box text: "CIRCUS BUGLER MECHANICAL TROMBONE PLAYING CLOWN ON STILTS." Box marks: T.P.S.

**Action:** Right arm goes up and down while attached to trombone. The trombone moves in and out of clown's mouth giving the appearance of moving a trombone slide. Clown sways back and forth, moving at the waist.

# CIRCUS CLOWN

Windup TIN LITHO CLOWN ON UNICYCLE WITH MOVING ARMS and fixed key.
Toy Marks: Japan (rear left coat tail).
Size: H: 5.5in W: 2in D: 4in (H: 14cm W: 5cm D: 10cm)
Introduced: c.1957. Est. quantity: 10,000 pieces. Scarcity rating: 4.
*Fine to Excellent (C6 to C8): $235-$300, Mint (C10): $375.*

Box text: "MECHANICAL CIRCUS CLOWN."
*Photo courtesy of New England Auction Gallery.*

**Action:** Clown pedals forward on unicycle, spins around and moves forward again in different direction while arms swing up and down.

# CIRCUS CYCLIST

Windup CLOWN ON HIGH WHEELED TRICYCLE WITH CIRCUS IMAGES ON BELL and fixed key. Tin with satin like cloth clown outfit.
Toy Marks: T.P.S. (top of bell).
Size: H: 6.75in W: 2in L: 4.75in (H: 17cm W: 5cm L: 12cm)
Introduced: c.1956. Est. quantity: 12,000 pieces. Scarcity rating: 4.
*Fine to Excellent (C6 to C8): $450-$600, Mint (C10): $750.*

Box text: "MECHANICAL CIRCUS CYCLIST." Box marks: T.P.S. *Photo courtesy of New England Auction Gallery.*

**Action:** Clown pedals as cycle with ringing bell, moves in a wide circle.

Circus lithography on bell.

# CIRCUS ELEPHANT

Windup ELEPHANT WITH RIDER BALANCING ON SPINNING BALL with fixed key. Tin.
Toy Marks: Mitsuhashi (rear of ball near base).
Size: H: 6.25in W: 2in L: 4.75in (H: 16cm W: 5cm L: 12cm)
Introduced: c.1956. Est. quantity: 10,000 pieces. Scarcity rating: 4.
*Fine to Excellent (C6 to C8): $185-$235, Mint (C10): $300.*

Box text: "CIRCUS ELEPHANT," "M-818." Box marks: Mitsuhashi.

**Action:** Elephant, with bell on nose and rider on top, moves head up and down while going forward on spinning ball. Toy turns around multiple times via lever extending from base then goes in different direction.

# CIRCUS MONKEY

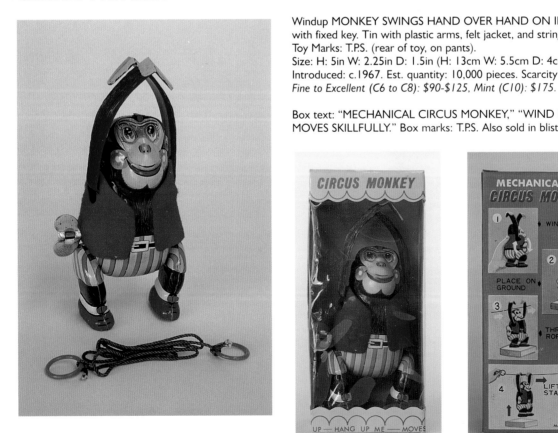

Windup MONKEY SWINGS HAND OVER HAND ON INCLUDED 25" STRING, with fixed key. Tin with plastic arms, felt jacket, and string.
Toy Marks: T.P.S. (rear of toy, on pants).
Size: H: 5in W: 2.25in D: 1.5in (H: 13cm W: 5.5cm D: 4cm)
Introduced: c.1967. Est. quantity: 10,000 pieces. Scarcity rating: 4.
*Fine to Excellent (C6 to C8): $90-$125, Mint (C10): $175.*

Box text: "MECHANICAL CIRCUS MONKEY," "WIND ME UP * HANG UP ME * MOVES SKILLFULLY." Box marks: T.P.S. Also sold in blister type package.

Box side panel showing instructions and toy action.

**Action:** Hanging the monkey on string activates start mechanism. Monkey moves forward on string with hand over hand action.

## CIRCUS PARADE
## a.k.a. ELEPHANT CIRCUS PARADE

Windup ELEPHANT PULLS THREE PERFORMING CLOWNS IN CIRCUS PARADE, with fixed key. Tin.
Toy Marks: T.P.S. (right rear elephant leg).
Size: L: 13.5in H: 3.5in W: 2.75in (L: 34cm H: 9cm W: 7cm)
Introduced: c.1959. Est. quantity: 100,000 pieces. Scarcity rating: 2.
*Fine to Excellent (C6 to C8): $125-$200, Mint (C10): $275.*

Box text: "MECHANICAL CIRCUS PARADE" or "MECHANICAL ELEPHANT CIRCUS PARADE," "CRAGSTAN 11669." Box marks: T.P.S., Cragstan. *Brynne and Scott Shaw Collection.*

**Action:** Elephant pulls clowns connected by rods. Wheel friction causes two clowns to spin on balls and one clown to spin inside hoop. *Brynne and Scott Shaw Collection.*

Circus graphics on box end panel.
*Brynne and Scott Shaw Collection.*

# CIRCUS SEAL

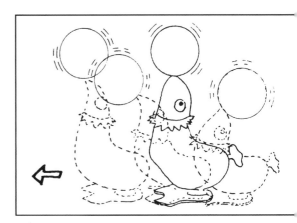

Windup PLUSH COVERED SEAL SPINNING BALL with fixed key. Flock covered tin with cloth collar and plastic ball. *Photo courtesy of New England Auction Gallery.*
Size: H: 6.75in L: 4in W: 3.25in (H: 17cm L: 10cm W: 8cm)
Introduced: c.1967. Est. quantity: 12,000 pieces. Scarcity rating: 4.
*Fine to Excellent (C6 to C8): $75-$90, Mint (C10): $110.*

Box text: "MECHANICAL CIRCUS SEAL." Box marks: T.P.S. *Photo courtesy of New England Auction Gallery.*

**Action:** Seal rocks back and forth while spinning ball. Move forward through vibrating action.

# CLIMBING LINESMAN

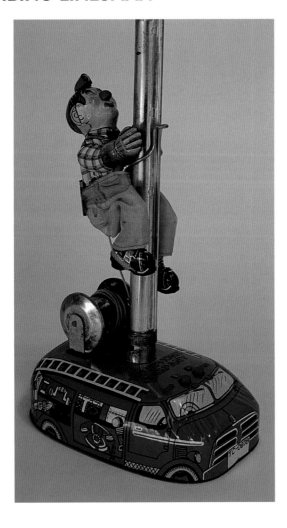

Battery operated TELEPHONE COMPANY TRUCK WITH 3 SECTION POLE & CLIMBING LINESMAN. Tin.
Toy Marks: T.P.S. (right rear truck door).
Size: H: 24in W: 4in D: 5.75in (H: 61cm W: 10cm D: 14.5cm)
Introduced: c.1957. Est. quantity: 6,000 pieces. Scarcity rating: 4.5
*Fine to Excellent (C6 to C8): $400-$525, Mint (C10): $675.*

**Action:** Up-Down switch on truck causes linesman, with hat light, to climb up and down pole. Legs and arms move in climbing motion . Cable is let out or taken up from reel.

Box text: "BATTERY OPERATED CLIMBING LINESMAN," "REPAIRMAN CLIMBS UP & DOWN * CABLE EASES OUT AND REWINDS AUTOMATI-CALLY * LAMP ON REPAIRMAN STAYS LIT WHILE IN OPERATION." Box marks: T.P.S., HTC. AHI mark has been covered over with paper label.

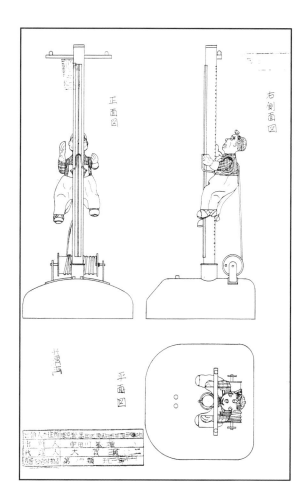

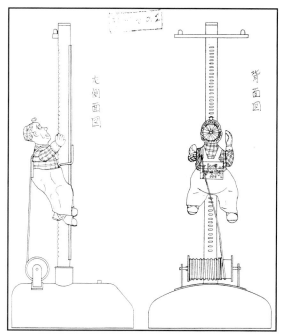

Outline Drawing. Japanese Patent # 138123 granted 1958.
*Courtesy of Toplay (T.P.S.) Ltd.*

# CLIMBING PIRATE

STRING CLIMBING PIRATE. Tin with felt jacket and satin like pants and head scarf.
Toy Marks: T.P.S. (left side near belt).
Size: H: 7in W: 2.25in D: 2.25in (H: 18cm W: 6cm D: 6cm)
Introduced: c.1962. Est. quantity: 12,000 pieces. Scarcity rating: 3.5.
*Fine to Excellent (C6 to C8): $140-$180, Mint (C10): $225.*

Box text: "CLIMBING PIRATE." Box marks: T.P.S. *Don Hultzman Collection.*

**Action:** Pulling string taut causes pirate to climb upward, relaxing string causes pirate to climb downward. String has metal loop at both ends. *Don Hultzman Collection.*

## CLIMBO THE CLIMBING CLOWN

STRING CLIMBING CLOWN. Tin with felt jacket and cloth pants.
Toy Marks: T.P.S. (top of right shoe).
Size: H: 6.75 W: 2.75in D: 2.25in (H: 17cm W: 7cm D: 6cm)
Introduced: c.1962. Est. quantity: 12,000 pieces. Scarcity rating: 3.5.
*Fine to Excellent (C6 to C8): $180-$235, Mint (C10): $310.*

Box text: "CLIMBO the CLIMBING CLOWN,"
"PULL MY RING AND SEE ME CLIMB *LET GO
AND I COME BACK DOWN * PAT.PEND." Box
marks: T.P.S., Franconia.

**Action:** Pulling string taut causes figure to climb
upward, relaxing string causes figure to climb
downward. String has metal loop at both ends.

## CLOWN JALOPY CYCLE

Friction CLOWN RIDING BREAK APART MOTORCYCLE. Tin. Known
variations: Box spelling "CROWN JALOPY CYCLE." *Photo courtesy of Carl
Johnson and Karen Delp.*
Toy Marks: T.P.S. (left side rear fender).
Size: L: 8.75in H: 6.25in W: 2.25in (L: 22cm H: 16cm W: 5.5cm)
Introduced: c.1963. Est. quantity: 12,000 pieces. Scarcity rating: 4.
*Fine to Excellent (C6 to C8): $325-$425, Mint (C10): $550.*

Box text: "NEW ACTION TOY," "CLOWN JALOPY CYCLE,"
"FRICTION POWERED." Box marks: T.P.S. *Photo courtesy of Carl
Johnson and Karen Delp.*

**Action:** Friction powered motorcycle moves forward until front
fender hits object, causing latch to spring open front part of the
cycle. Clown's head pops up and down.

# CLOWN JUGGLER WITH BALL

Windup CLOWN TWIRLING BALL BALANCED ON NOSE, with fixed key. *Japan Toys Museum Foundation.*
**Action:** Clown with rotating arms (right one holds stick) hits ball balanced on his nose and causes it to spin.
Toy Marks: T.P.S. (rear of right leg near base).
Size: H: 6in W: 2.75in D: 2in (H: 15cm W: 7cm D: 5cm)
Introduced: c.1958. Est. quantity: 6,000 pieces. Scarcity rating: 4.5.
*Fine to Excellent (C6 to C8): $300-$400, Mint (C10): $525.*

# CLOWN JUGGLER WITH MONKEY

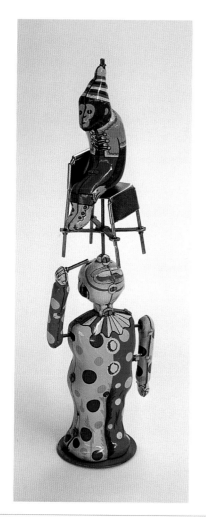

Windup CLOWN TWIRLING MONKEY ON CHAIR BALANCED ON NOSE, with fixed key. Tin.
Toy Marks: T.P.S. (rear of right leg near base).
Size: H: 9.25in W: 2.75in D: 2in (H: 23.5cm W: 7cm D: 5cm)
Introduced: c.1959. Est. quantity: 10,000 pieces. Scarcity rating: 4.
*Fine to Excellent (C6 to C8): $500-$675, Mint (C10): $900.*

Box text: "MECHANICAL CLOWN JUGGLER WITH MONKEY."
Box marks: T.P.S.

**Action:** Clown's arms rotate. The right arm has a stick which hits chair leg causing monkey to spin while balanced on clown's nose. Chair is attached by rod.

# CLOWN MAKING LION JUMP THRU FLAMING HOOP

Windup LION JUMPS THROUGH FLAMING HOOP HELD BY CLOWN, with fixed key. Tin with cloth flames and rubber tail on lion.
Toy Marks: T.P.S. (rear of clown near base), Cragstan (front of clown near base).
Size: L: 7.5in H: 5.5in W: 3in (L: 19cm H: 14cm W: 8cm)
Introduced: c.1960. Est. quantity: 15,000 pieces. Scarcity rating: 3.5.
*Fine to Excellent (C6 to C8): $175-$225, Mint (C10): $275.*

Box text: "Clown Making THE Lion Jump Thru THE Flaming Hoop," "AN ANIMATED WIND-UP TOY * 12630." Box marks: T.P.S., Cragstan.

**Action:** Clown rotates around to follow the lion that runs around in a circle and jumps through the flaming hoop attached to base. Left arm, with trainer's stick, moves up and down as if commanding the lion.

# CLOWN ON ROLLER SKATE

Windup HOBO CLOWN ON ROLLER SKATES, with fixed key. Tin with satin like pants. Known variations: Red haired version seen in Japan. Handmade larger version (100cm x 190cm x 175cm) as discussed in Introduction.
Toy Marks: T.P.S. (back, left side).
Size: H: 6.25in W: 3.25in D: 4in (H: 16cm W: 8cm D: 10cm)
Introduced: c.1954. Est. quantity: 1,000,000 pieces. Scarcity rating: 1.
*Fine to Excellent (C6 to C8): $175-$250, Mint (C10): $300.*

Box text: "MECHANICAL CLOWN ON ROLLER SKATE." Box marks: T.P.S., HTC.

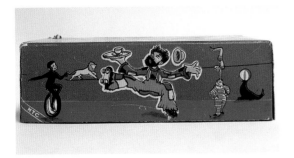

**Action:** Clown skates on wheeled left foot with right leg pushing off, causing body to dip in realistic skating motion.

Box end panel with circus graphics.

Cutaway and outline drawing for "Clown on Roller Skate."
*Courtesy of Toplay (T.P.S.) Ltd.*

# CLOWN ON ROLLER SKATE — WHITE FACE

Windup ROLLER SKATING CLOWN WITH WHITE FACE AND FLESH COLORED HANDS, with fixed key. Tin with satin like pants and felt jacket.
Toy Marks: None.
Box text: "MECHANICAL CLOWN ON ROLLER SKATE"
**Action:** Clown skates on wheeled left foot with right leg pushing off, causing body to dip in realistic skating motion.
Size: H: 6.25in W: 3.25in D: 4in (H: 16cm W: 8cm D: 10cm)
Introduced: c.1956. Est. quantity: 6,000 pieces. Scarcity rating: 4.5.
*Fine to Excellent (C6 to C8): $600-$750, Mint (C10): $1,000.*

# CLOWN TRAINER AND HIS ACROBATIC DOG
## a.k.a. CLEO CLOWN THE DOGS

Windup POODLE TYPE DOG JUMPING THROUGH FLAMING HOOP WITH CLOWN TRAINER, with fixed key. Tin with felt flame and rubber tail on dog. Also sold as "Cleo Clown the Dogs" with Rosko and T.P.S. marks on box.
Toy Marks: T.P.S. (below key).
Size: H: 5.5in L: 7.5in W: 6in (L: 19cm H: 14cm W: 8cm)
Introduced: c.1960. Est. quantity: 30,000 pieces. Scarcity rating: 3.
*Fine to Excellent (C6 to C8): $150-$200, Mint (C10): $250.*

Box text: "Mechanical Clown Trainer & His Acrobatic Dog" or "Cleo Clown the Dogs." "WATCH HER RUNS & JUMPS THROUGH FIRE RING." Box marks: T.P.S.

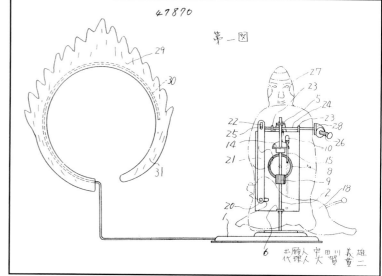

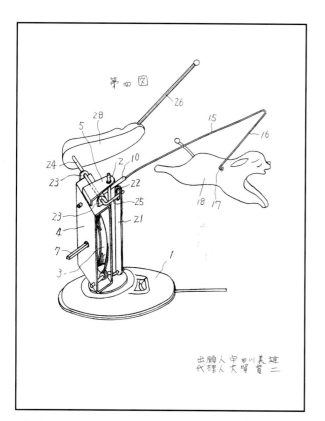

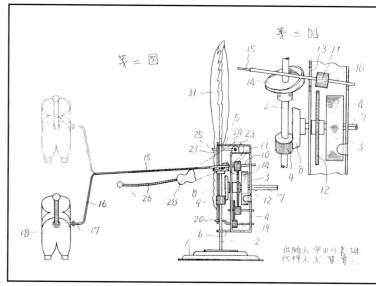

Clockwork mechanism of "Clown Trainer and his Acrobatic Dog."
**Action:** Clown rotates around to follow the dog that runs around in a circle and jumps through the flaming hoop attached to base. Left arm, with trainer's stick, moves up and down as if commanding the dog.

# CLOWN'S POPCORN TRUCK

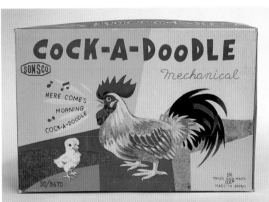

Battery operated TRUCK WITH CLOWN DRIVER AND STYROFOAM BALLS BLOWING IN DRUM. Tin with plastic wheels, head, and popcorn drum. *Don Hultzman Collection.*
Toy Marks: T.P.S. (top rear of wagon).
Size: L: 6in H: 6in W: 4in (L: 15cm H: 15.5cm W: 10cm)
Introduced: c.1970. Est. quantity: 15,000 pieces. Scarcity rating: 3.5.
*Fine to Excellent (C6 to C8): $150-$200, Mint (C10): $275.*

Box text: "BATTERY OPERATED COLORFUL BALL BLOWING TOY," "CLOWN'S POPCORN TRUCK."
Box marks: T.P.S. *Don Hultzman Collection*

**Action:** Clown's mouth opens and closes as truck moves in bump and go action with engine sound. Popcorn like balls are blown around inside clear drum. *Don Hultzman Collection.*

# COCK-A-DOODLE

Windup CROWING ROOSTER WITH FEATHERS and fixed key. Tin with fragile plastic feathers. Known variations: Also sold as "Jumping Cock" with spring jumping mechanism instead of crowing.
Toy Marks: T.P.S. (bottom left side).
Size: L: 7.5in H: 4.25in W: 3.25in (L: 19cm H: 11cm W: 8cm)
Introduced: c.1965. Est. quantity: 20,000 pieces. Scarcity rating: 3.5.
*Fine to Excellent (C6 to C8): $85-$100, Mint (C10): $150.*

Box text: "Mechanical COCK-A-DOODLE," "HERE COMES MORNING * COCK-A-DOODLE * 30/9670." Box marks: T.P.S., Sonsco.

**Action:** Rooster rocks back and forth and vibrates while crowing sounds come from voice box.

## COLORFUL BALL BLOWING TRUCKS

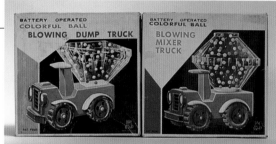

Battery operated TRUCK WITH MULTICOLOR STYROFOAM BALLS BLOWING IN RECTANGULAR CONTAINER OR MIXING DRUM. Plastic with Styrofoam balls.
Toy Marks: T.P.S. (molded in front bumper).
Size: L: 6.75in H: 6in W: 4in (L: 17cm H: 15cm W: 10cm)
Introduced: c.1970. Est. quantity: 15,000 pieces each. Scarcity rating: 3.5.
*Fine to Excellent (C6 to C8): $50-$65, Mint (C10): $85.*

Box text: "BATTERY OPERATED COLORFUL BALL BLOWING DUMP / MIXER TRUCK," "NON-STOP ACTION * ENGINE SOUND * THE COLORFUL BALLS ARE BLOWN UP INSIDE TRANSPARENT VAN #8084." Box marks: T.P.S.
**Action:** Trucks moves in bump and go action with engine sound, as balls are blown around inside clear container.

## COMBAT TANK ON BATTLE FRONT

Windup 6cm (2.25in) TANK ON PLATFORM BASE WITH TUNNEL AND MILITARY BATTLE SCENES, with separate key. Tin.
Toy Marks: T.P.S. (base, lower right corner).
Size: 15.25in x 6.75in platform (38.5cm x 17cm platform)
Introduced: c.1963. Est. quantity: 10,000 pieces. Scarcity rating: 4.
*Fine to Excellent (C6 to C8): $125-$160, Mint (C10): $200.*

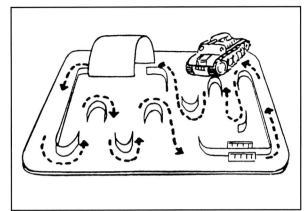

Box text: "MECHANICAL COMBAT TANK ON BATTLE FRONT," "HOOK METAL KNOB UNDER TANK TO ROAD GUARD." Box marks: T.P.S.

**Action:** Windup US Army tank goes around platform and through tunnel hooked on zigzag guide rail.

# COMICAL CLARA

Windup WIGGLING GIRL WITH POPPING EYES and fixed key. Tin.
Toy Marks: T.P.S. (back side of dress).
Size: H: 5in W: 2.75in D: 2.75in (H: 13cm W: 7cm D: 7cm)
Introduced: c.1967. Est. quantity: 12,000 pieces. Scarcity rating: 3. Note: A 1998 warehouse find of old stock has increased the availability of this toy
*Fine to Excellent (C6 to C8): $300-$375, Mint (C10): $450.*

Box text: "MECHANICAL COMICAL CLARA," "THE CUTE KIDS WITH THE COMICAL EYES AND THE WIG WAG WALK." Box marks: T.P.S.

**Action:** Girl wiggles back and forth causing arms to move like she was walking, while her eyes pop in and out in a comical fashion.

# CONNY ISLAND SCOOTER

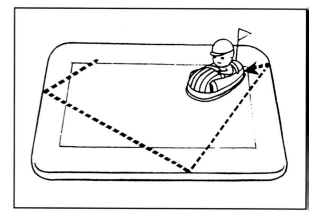

Windup 7cm (2.75in) BUMPER CAR WITH FLAG, ON AMUSEMENT PARK PLATFORM BASE with fixed key. Tin with vinyl head on bumper car.
Toy Marks: T.P.S. (base, lower right corner).
Size: 10.25in x 10.25in platform (26cm x 26cm platform)
Introduced: c.1965. Est. quantity: 20,000 pieces. Scarcity rating: 3.5.
*Fine to Excellent (C6 to C8): $175-$200, Mint (C10): $250.*

Box text: "NEW ACTION TOY," "CONNY ISLAND SCOOTER," "MECHANICAL-MYSTERY NON STOP ACTION." Box marks: T.P.S. While the correct spelling was Coney Island Scooter, "Conny Island Scooter" appears to be the only version produced.

**Action:** Windup bumper car with mystery bump and go action on platform. Car stays within the confines of the flat rail on base via knob protruding from underside of car.

# COW HOUSE WITH VOICE

Windup COW THAT MOVES IN AND OUT OF BARN LIKE HOUSE, with fixed key. Tin.
Box text: "Crank Action Cow House with Voice."
**Action:** Turning crank causes cow to move out of barn to feed. Cow bellows and moves back inside.
Size: L: 9.5in (L: 24cm)
Introduced: c.1965. Est. quantity: 6,000 pieces. Scarcity rating: 5.
*Fine to Excellent (C6 to C8): $175-$250, Mint (C10): $350.*

# CRANE TRUCK

Friction STEERABLE THREE AXLE EAGLE CRANE TRUCK WITH WORKING BUCKET. Tin with rubber wheels. Known variations: Later version for Japanese market with lithographed windows on crane and truck cab instead of open windows, Japanese writing on cab door and no interior detail. Toy Marks: Marusan and Hayashi (truck cab dashboard). Size: L: 15.75in H: 9.5in W: 4.25in (L: 40cm H: 24cm W: 11cm) Introduced: c.1970. Est. quantity: 30,000 pieces. Scarcity rating: 4.5.
*Fine to Excellent (C6 to C8): $250-$300, Mint (C10): $400.*

**Action:** Rear wheel friction powered truck with hand crank controlling bucket payout and opening. Crane rotates 360°. Side lever adjusts boom height. Front wheels can be positioned for direction. Crane base slides off truck bed.

# DANCING CLOWN

Windup PLASTIC CLOWN THAT DANCES IN CIRCLES, with fixed key. Plastic.
Toy Marks: T.P.S. (paper label on bottom).
Size: H: 6.25in W: 4.75in D: 2in (H: 16cm W: 12cm D: 5cm)
Introduced: c.1981. Est. quantity: 30,000 pieces. Scarcity rating: 4.
*Fine to Excellent (C6 to C8): $45-$55, Mint (C10): $75.*

Box text: "MECHANICAL DANCING CLOWN" (In Japanese and English). Box marks: T.P.S.

**Action:** Head, arms, and feet move as clown goes sideways, then turns in circles via lever which extends from base to raise front wheel from surface.

79

# DANCING COUPLE

Windup GIRL AND BOY BALLROOM DANCING IN FORMAL ATTIRE, with fixed key. Tin with vinyl heads. Known variations: Dark haired boy, add 15-20%.
Toy Marks: Hishimo (lower right side of girl's gown).
Size: H: 5.5in W: 2.5in D: 2.75in (H: 14cm W: 6cm D: 7cm)
Introduced: c.1960. Est. quantity: 12,000 pieces. Scarcity rating: 3.5.
*Fine to Excellent (C6 to C8): $85-$120, Mint (C10): $150.*

Box text: "MECHANICAL 'DANCING COUPLE.'" Box marks: T.P.S., Hishimo, MEGO.

**Action:** Couple moves across the floor, spins around twice via lever extending from base, and repeats this dancing action.

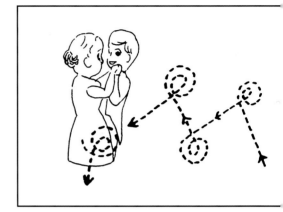

Ballroom dancing box graphics.

# Dax HONDA MOTOR CYCLE

Windup Dax 70 HONDA MOTOR CYCLE WITH RIDER and fixed key. Plastic with rubber tires and vinyl rider. Known variations: Red, blue, or brown.
Toy Marks: Japan (left side).
Size: L: 5.5in H: 5.5in W: 2in (L: 14cm H: 14cm W: 5cm)
Introduced: c.1975. Est. quantity: 20,000 pieces. Scarcity rating: 4.
*Fine to Excellent (C6 to C8): $75-$100, Mint (C10): $125.*

Box text: "MECHANICAL Dax HONDA MOTOR CYCLE NO. 4044." (In Japanese and English) Box marks: T.P.S.

**Action:** Motorcycle with engine noise rides on front wheel and two drive wheels for stability. Gear ratio allows for fast running. Front wheel is adjustable to determine direction. Rider is detachable.

# DOG HOUSE WITH VOICE

Windup DOG THAT MOVES IN AND OUT OF DOG HOUSE, with fixed key.
Size: L: 8.75in (L: 22cm)
Introduced: c.1965. Est. quantity: 6,000 pieces. Scarcity rating: 5.
*Fine to Excellent (C6 to C8): $175-$250, Mint (C10): $350.*

Box text: "CRANK ACTION DOG HOUSE WITH VOICE".

**Action:** Turning crank causes dog to move out of house to get bone. Dog barks and moves back inside.

# DREAM LAND BUS in MAGIC TUNNEL

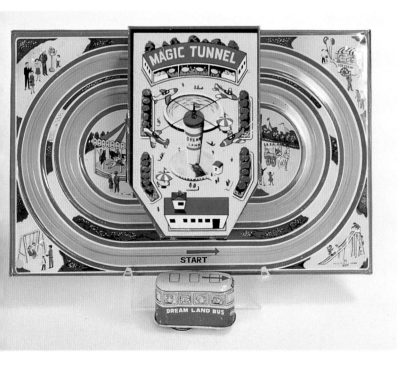

Windup 5cm (2in) DREAM LAND BUS ON PLATFORM BASE WITH AMUSEMENT PARK SCENE, with separate key. Tin.
Toy Marks: T.P.S. (lower right corner of base).
Size: 9.25in x 6.25in platform (23.5cm x 15.5cm platform)
Introduced: c.1961. Est. quantity: 70,000 pieces. Scarcity rating: 2.
*Fine to Excellent (C6 to C8): $95-$125, Mint (C10): $145.*

Box text: "MECHANICAL DREAM LAND BUS in MAGIC TUNNEL." Box marks: T.P.S.

**Action:** Bus runs on circular track on platform, and switches tracks in magic tunnel via guide wheel on bus and groove in track.

## DREAM SUPER EXPRESS — "THE HIKARI"

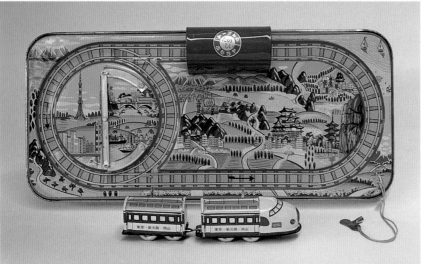

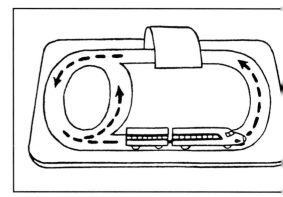

Windup 17cm (6.75in) TWO CAR BULLET TRAIN (SHINKANSEN) ON PLATFORM BASE WITH SCENES OF JAPAN'S MAJOR CITIES AND COUNTRYSIDE, with separate key. Tin.
Toy Marks: T.P.S. (lower right hand corner).
Size: L: 15.25in W: 6.75in H: 2in (L: 38.5cm W: 17cm H: 5cm)
Introduced: c.1964. Est. quantity: 30,000 pieces. Scarcity rating: 3.
*Fine to Excellent (C6 to C8): $100-$125, Mint (C10): $150.*

Box text: "MECHANICAL DREAM SUPER EXPRESS 'THE HIKARI' WITH TOKAIDO SCENERY BOARD" (In Japanese). Box marks: T.P.S.

**Action:** Bullet train travels around track and through tunnel on platform, in small or large circular pattern determined by automatic rail changing mechanism.

## DREAMLAND AIR PORT

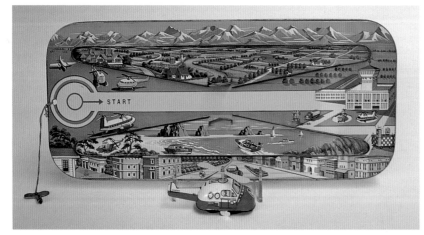

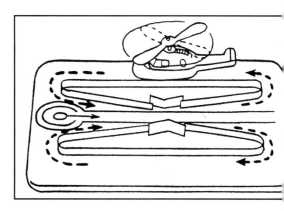

Windup 9cm (3.5in) HELICOPTER ON PLATFORM BASE WITH AIRFIELD AND COUNTRYSIDE SCENES, with separate key. Tin with plastic helicopter rotor blade.
Toy Marks: T.P.S. (bottom left side of base), Hirata (bottom center of base).
Size: 15.25in x 6.75in platform (38.5cm x 17cm platform)
Introduced: c.1965. Est. quantity: 20,000 pieces. Scarcity rating: 4.
*Fine to Excellent (C6 to C8): $125-$150, Mint (C10): $175.*

Box text: "DREAMLAND AIR PORT," "WIND UP MECHANISM." Box marks: T.P.S.

**Action:** Helicopter with spinning propeller goes around airfield platform in figure eight pattern while hooked on guide rail.

82

# DREAMLAND MIDGET CHOO CHOO

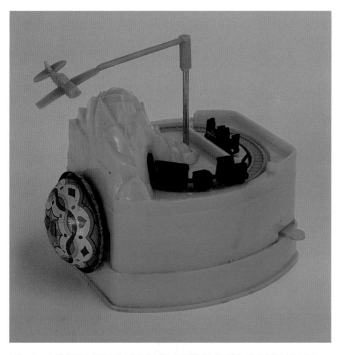

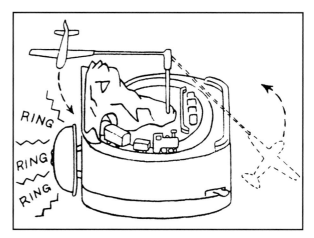

Windup MINIATURE TRAIN AND AIRPLANE CIRCLING MOUNTAIN WITH TUNNEL AND RINGING BELL, with fixed key. Plastic with tin bell.
Toy Marks: Japan (molded in bottom of base).
Size: L: 3.5in H: 3.25in W: 2.75in (L: 9cm H: 8.5cm W: 7cm)
Introduced: c.1967. Est. quantity: 50,000 pieces. Scarcity rating: 3.
*Fine to Excellent (C6 to C8): $45-$55, Mint (C10): $75.*

Box text: "DREAMLAND MIDGET CHOO CHOO W/BELL," "MECHANICAL PLASTIC ACTION TOY." Box marks: T.P.S.

**Action:** Small train and suspended airplane rotate around base on common axis as bell rings. Has start-stop switch.

# DREAMLAND WITH BELL

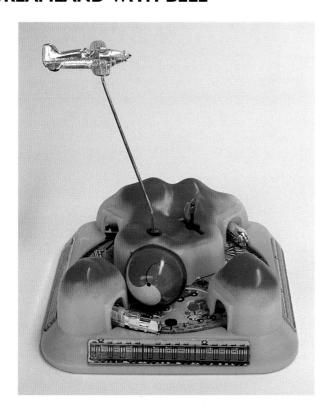

Windup AMUSEMENT PARK WITH ROTATING TRAINS AND PLANE, with fixed key. Plastic with tin amusement park, bell, bottom, and train plaques on each side. Known variations: Fixed or separate key, metal or plastic key. Taiwanese version from Soon Cheng Toys.
Toy Marks: None.
Size: W: 6in D: 6in H: 6in (W: 15cm D: 15cm H: 15cm)
Introduced: c.1980. Est. quantity: 100,000 pieces. Scarcity rating: 2.
*Fine to Excellent (C6 to C8): $35-$45, Mint (C10): $65.*

Box text: "MECHANICAL DREAMLAND WITH BELL," "LONG LASTING * NOT EASILY BROKEN * 3 TUNNELS * LARGE CIRCLE * USE SPECIAL SPRING" (In Japanese & English). Box marks: T.P.S.

**Action:** Two trains travel around amusement park with ringing bell. Plane on detachable rod flies around park.

# DRIVE TESTER

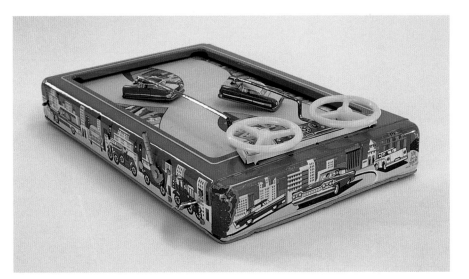

Windup DRIVING GAME WITH TWO STEERING WHEEL CON-TROLLED CARS ON MOVING HIGHWAY, with fixed key. Tin with paper highway and plastic steering wheels.
Toy Marks: T.P.S. (lower right corner of front side).
Size: L: 10.75in W: 6.75in H: 2in (L: 27cm W: 17cm H: 5cm)
Introduced: c.1961. Est. quantity: 10,000 pieces. Scarcity rating: 4.
*Fine to Excellent (C6 to C8): $150-$190, Mint (C10): $250.*

Box text: "DRIVE TESTER," "2 CARS EACH WITH CONTROL STEERING. MOVING REALISTIC HIGHWAY. RINGING BELL," "IT'S FUN AND THRILLING TO COMPETE WITH DAD. SEE WHO IS THE BETTER." Box marks: T.P.S.

**Action:** Start switch causes paper highway with country and city scenes to move over rolls. Objective is to drive two tin cars connected to steering wheels via rods, over the highway. Includes ringing bell.

# DUCK AMPHIBIOUS TAXI

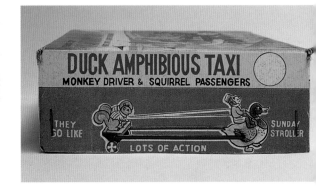

Windup DUCK WITH MONKEY DRIVER PULLING SQUIRRELS ON CART, with fixed key. Tin with rubber feet.
Toy Marks: T.P.S. (rear left side of duck).
Size: L: 6.75in H: 4.25in W: 2.5in (L: 17cm H: 11cm W: 6.5cm)
Introduced: c.1960. Est. quantity: 10,000 pieces. Scarcity rating: 4.5.
*Fine to Excellent (C6 to C8): $500-$650, Mint (C10): $850.*

Box text: "MECHANICAL TOY * DUCK AMPHIBIOUS TAXI," "MONKEY DRIVER & SQUIRREL PASSENGERS * THEY GO LIKE SUNDAY STROLLER * LOTS OF ACTION." Box marks: T.P.S.

**Action:** Duck walks as head and neck move back and forth. Monkey driver and squirrels are attached to duck's neck via reins and rock back and forth as neck moves.

Duck amphibious taxi crossing river.

## DUCK FAMILY PARADE

Windup MOTHER DUCK WITH THREE BABY DUCKS, with fixed key.
Tin with rubber wheels.
Toy Marks: T.P.S. (left rear tail of mother duck).
Size: L: 11.75in H: 3.25in W: 2.25in (L: 30cm H: 8.5cm W: 5.5cm)
Introduced: c.1959. Est. quantity: 50,000 pieces. Scarcity rating: 2.5.
*Fine to Excellent (C6 to C8): $95-$115, Mint (C10): $140.*

Box text: "MECHANICAL DUCK FAMILY PARADE." Box marks: T.P.S.,
Hikari.

**Action:** Mother duck moves mouth while pulling attached baby ducks on
wheels as they spin around.

## DUCK THE MAIL MAN

Windup DUCK (SIMILAR TO DONALD DUCK) DRESSED AS
MAILMAN CARRYING GEESE IN EACH ARM, with fixed key. Tin.
Toy Marks: T.P.S. (rear left side).
Size: H: 4.25in W: 4in D: 3.5in (H: 11cm W: 10cm D: 9cm)
Introduced: c.1960. Est. quantity: 6,000 pieces. Scarcity rating: 4.5.
*Fine to Excellent (C6 to C8): $475-$625, Mint (C10): $800.*

Box text: "DUCK THE MAILMAN," "TURN-O-GO ACTION *
WIND UP CLOCK MOVEMENT." Box marks: T.P.S.

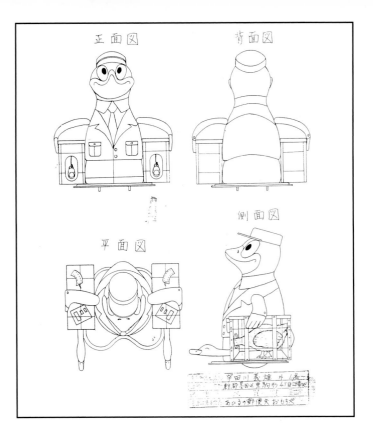

**Action:** Duck walks (on wheels) with geese heads sticking out of cases. He stops and makes about 1-1/2 turns before moving forward again. Lever extending from base raises one wheel from surface to accomplish turns.

"Duck The Mailman Drawing." Japanese Patent #202680 granted 1961. *Courtesy of Toplay (T.P.S.) Ltd.*

# DUNE BUGGY W/ SURF BOARD

Battery operated BLUE DUNE BUGGY WITH SURFBOARD, THAT DOES WHEELIES (FIXED FRONT WHEELS). Tin with plastic fenders, driver, and tires.
Toy Marks: T.P.S. (right passenger seat), Made in Japan (top of surfboard).
Size: L: 11in W: 5.5in H: 5.5in (L: 28cm W: 14cm H: 14cm)
Introduced: c.1971. Est. quantity: 10,000 pieces. Scarcity rating: 4.
*Fine to Excellent (C6 to C8): $100-$125, Mint (C10): $160.*

Box text: "BATTERY OPERATED DUNE BUGGY W/ SURF BOARD," "WHEEL STAND ACTION!! * IT DOES A REAL WHEEL STAND * TURNS AROUND-COMES DOWN * ROARING ENGINE." Box marks: T.P.S.

**Action:** Buggy goes forward. Lever with fifth wheel drops down causing front wheels to raise and spin buggy around. After lever retracts, the buggy goes forward again before repeating actions. Fixed front wheels.

Windup POOCH WITH GRADUATION CAP, NODS ANSWERS TO MATH PROBLEMS. Fixed key. Tin with cloth collar.
Toy Marks: T.P.S. (right side near dial).
Size: H: 4in W: 2.25in D: 2in (H: 10cm W: 6cm D: 5cm)
Introduced: c.1967. Est. quantity: 20,000 pieces. Scarcity rating: 3.
*Fine to Excellent (C6 to C8): $65-$85, Mint (C10): $110.*

Box text: "Magic Trick Performing EDUCATIONAL PET POOCH," "INSTRUCTIVE EDUCATIONAL PLAY-TOY AND MAGIC TRICK PERFORMER." Box marks: Shackman. Also seen with no marks .

**Action:** Turn dial underneath toy and dog nods number selected.

Box panel with instructions and action drawing.

# F. D. LADDER TRUCK

Windup 7cm (2.75in) LADDER TRUCK ON OPERATION FIRE BRIGADE PLATFORM BASE, with fixed key. Tin.
Toy Marks: T.P.S. (lower right corner of base).
Size: 10.25in x 10.25in platform (26 cm x 26cm platform)
Introduced: c.1965. Est. quantity: 20,000 pieces. Scarcity rating: 3.5.
*Fine to Excellent (C6 to C8): $115-$140, Mint (C10): $175.*

Box text: "BUMP GO NON STOP ACTION F.D. LADDER TRUCK." Box marks: T.P.S.

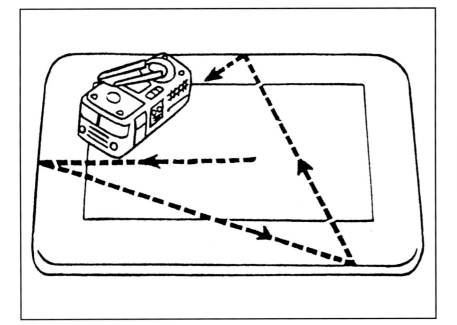

**Action:** Fire Truck moves within flat rail around platform base with bump and go action initiated by knob protruding from bottom of truck. Key slot in base.

## FAMILY GIRAFFE LOCO

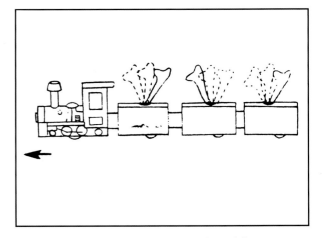

Windup ARTICULATED 3 CAR TRAIN, EACH WITH GIRAFFE'S NECK STICKING THROUGH ROOF. CARS ARE MARKED "DADDY, MAMMY & KIDDY," with fixed key. Tin with rubber wheels. *Brynne and Scott Shaw Collection.*
Toy Marks: Made in Japan (back of each car and loco).
Size: L: 11.25in H: 3in W: 1.25in (L: 28.5cm H: 7.5cm W: 3cm)
Introduced: c.1964. Est. quantity: 20,000 pieces. Scarcity rating: 4.
*Fine to Excellent (C6 to C8): $175-$250, Mint (C10): $350.*

Box text: "MECHANICAL FAMILY GIRAFFE LOCO." Box marks: T.P.S. *Brynne and Scott Shaw Collection.*

**Action:** Train travels forward and jointed cars follow in zigzag fashion. As cars travel, giraffes swings their necks back and forth via slotted connection to car axles.

# FERRIS WHEEL TRUCK

Battery operated FERRIS WHEEL TRUCK WITH MULTICOLOR STYROFOAM BALLS BLOWING IN ENCLOSURE. Plastic with Styrofoam balls and tin Ferris wheel.
Toy Marks: T.P.S. (molded in front bumper).
Size: L: 7in H: 7.5 W: 4in (L: 18cm H: 19cm W: 10cm)
Introduced: c.1970. Est. quantity: 10,000 pieces. Scarcity rating: 4.
*Fine to Excellent (C6 to C8): $75-$100, Mint (C10): $125.*

Box text: "BATTERY OPERATED COLORFUL BALL BLOWING TOY," "FERRIS WHEEL TRUCK * NON-STOP ACTION * ENGINE SOUND * THREE PRETTY PIGS ROUND. THE COLORFUL BALLS ARE BLOWN UP INSIDE TRANSPARENT DRUM." Box marks: T.P.S.

**Action:** Truck moves in bump and go action with engine sound as balls are blown around inside enclosure. Ferris wheel with pigs turns with air movement.

# FIGURE 8 HIGHWAY

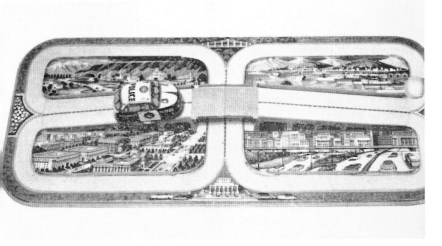

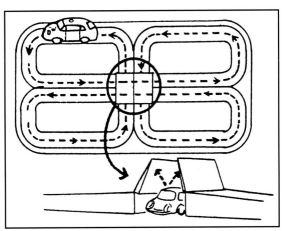

Windup 6cm (2.25in) POLICE CAR ON OVER AND UNDER FIGURE EIGHT PLATFORM BASE WITH CITY AND COUNTRY SCENES, with separate key. Tin.
Toy Marks: T.P.S.
Size: L: 15.25in W: 6.75in H: 2in (L: 38.5cm W: 17cm H: 5cm)
Introduced: c.1965. Est. quantity: 12,000 pieces. Scarcity rating: 4.
*Fine to Excellent (C6 to C8): $125-$150, Mint (C10): $175.*

Box text: "MECHANICAL FIGURE 8 HIGHWAY WITH CROSS OVER BRIDGE"

**Action:** Windup car goes around platform hooked on rail. Track elevates and crosses over itself via opening bridge, in figure eight fashion.

# FIREMAN V

Battery operated 10cm (4in) CLIMBING FIREMAN WITH 5 SECTION 123cm (48in) HIGH LADDER. Plastic.
Toy Marks: DY (molded on battery cover).
Size: H: 48.5in W: 6.25in D: 9in (H: 125cm W: 16cm D: 23cm)
Introduced: c.1984. Est. quantity: 100,000 pieces. Scarcity rating: 2.5.
*Fine to Excellent (C6 to C8): $45-$55, Mint (C10): $65.*

Box text: "BATTERY OPERATED FIRE MAN V," "THIS INTREPID FIREFIGHTER AT A FLICK OF THE CONTROL BUTTON, SCALES THE 123 cm HIGH LADDER ONE STEADY STEP AT A TIME, AND THEN RETURNS. PRODUCED UNDER LICENCE BY TOPLAY (T.P.S.) LTD. JAPAN 1984 * MADE IN TAIWAN." Box marks: DY, T.P.S.

Fireman starting his long climb.

**Action:** Motorized fireman with hose climbs five section ladder one step at a time. Upon reaching top, he pushes hose nozzles forward as if shooting water onto a building before returning one step at a time.

# FISHING BEAR

Windup BEAR WITH FISH JUMPING IN AND OUT OF NET, with fixed key. Tin with mesh net and rubber ears. Known variations: Black or brown ears.
Toy Marks: T.P.S. (base, front right).
Size: H: 7.25in W: 2.5in L: 4.25in (H: 18.5cm W: 6.5cm L: 11cm)
Introduced: c.1957. Est. quantity: 20,000 pieces.
Scarcity rating: 3.5.
*Fine to Excellent (C6 to C8): $175-$220, Mint (C10): $275.*

Box text: "MECHANICAL FISHING BEAR." Box marks: T.P.S.

**Action:** Bear moves net up and down with right hand, causing fish to jump on fishing pole line held in left hand.

# FISHING MONKEY ON WHALES

Windup MONKEY RIDING AND FISHING ON WHALE WITH 2 FISH, with fixed key. Tin.
Toy Marks: T.P.S. (right side of whale tail).
Size: L: 9in H: 4in W: 2in (L: 23cm H: 10cm W: 5cm)
Introduced: c.1957. Est. quantity: 6,000 pieces.
Scarcity rating: 4. Note: A 1999 warehouse find has increased the availability of this toy.
*Fine to Excellent (C6 to C8): $225-$300, Mint (C10): $375.*

Box text: "MECHANICAL FISHING MONKEY ON WHALES," "WIND-UP AUTO-PILOT ACTION MOVES BACK AND FORTH." Box marks: T.P.S., Rosko paper label.

**Action:** Whale moves forwards then backwards while monkey rocks back and forth as if trying to reel in fish with moving fins.

# FLASH SPACE PATROL

Battery operated SPACE PATROL, Z-206 SPACE SHIP WITH 3 BLADED ROTOR. Tin with plastic rotor blade and lights.
Toy Marks: T.P.S. (right rear of space ship).
Size: L: 7.75in H: 4.75in W: 4.75in (L: 20cm H: 12cm W: 12cm)
Introduced: c.1968. Est. quantity: 30,000 pieces. Scarcity rating: 3.5.
*Fine to Excellent (C6 to C8): $200-$250, Mint (C10): $325.*

Box text: "BATTERY OPERATED FLASH SPACE PATROL," "MYSTERY ACTION * FLASHING LIGHT FROM EYES * WHIRLING ROTOR BLADE * REALISTIC ENGINE SOUND." Box marks: T.P.S.
**Action:** Bump and go action with flashing multicolor lights. Rotor blade spins and engine makes popping sound.

# FLYING AIR CAR a.k.a. HOVER CRAFT

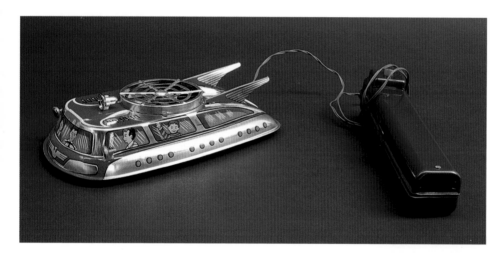

Battery operated RED AND SILVER FLOATING SPACESHIP TYPE CAR, WITH REMOTE BATTERY CONTROL. Aluminum with plastic lights and tin battery holder.
Toy Marks: T.P.S. (right rear side).
Size: L: 7.75in W: 4.75in H: 2.25in (L: 20cm W: 12cm H: 5.5cm)
Introduced: c.1966. Est. quantity: 30,000 pieces.
Scarcity rating: 4.
*Fine to Excellent (C6 to C8): $375-$500, Mint (C10): $675.*

Known variation: Early production had rotating plastic antenna and multi colored air moving blades (add 15%).

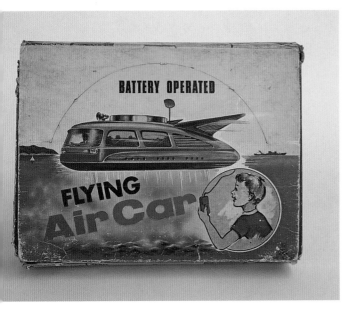

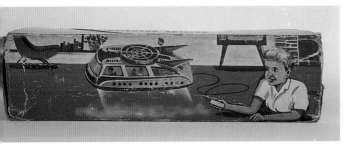

Box text: "BATTERY OPERATED FLYING Air Car" or "BATTERY OPERATED HOVER CRAFT." Box marks: T.P.S.

**Action:** Flying car rises, hovers, and darts above the surface, as rotating blades inside create air movement to lift the lightweight aluminum body off the ground.

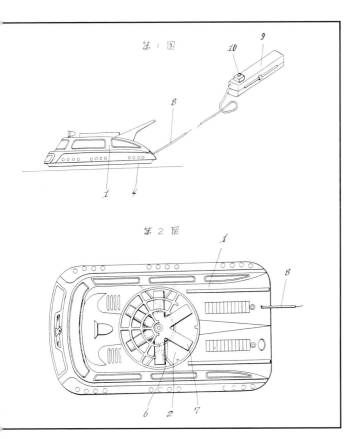

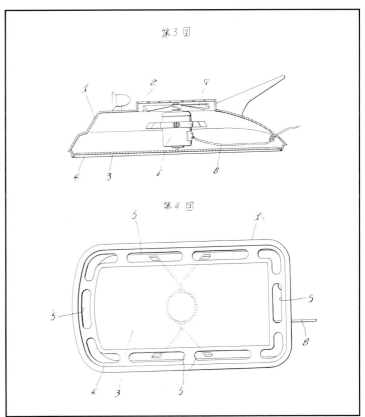

Outline drawing showing internal air flow mechanism. *Courtesy of Toplay (T.P.S.) Ltd.*

# FLYING BIRDS WITH VOICE

Windup TWO BIRDS FLYING OVER HOUSE AND FENCE ON CIRCULAR BASE WITH COUNTRY SCENE, with fixed key. Tin. Toy Marks: T.P.S. (side of base).
Size: H: 7.5in Base Diameter: 4.25in (H: 19cm Base Diameter: 11cm) Introduced: c.1965. Est. quantity: 12,000 pieces. Scarcity rating: 3.5.
*Fine to Excellent (C6 to C8): $150-$175, Mint (C10): $225.*

Box text: "MECHANICAL FLYING BIRDS WITH VOICE." Box marks: T.P.S. *Don Hulltzman Collection.*

**Action:** Two birds are attached to separate wires and fly in a circular pattern while making chirping sounds from voice box in base.

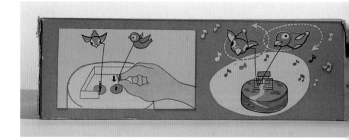

Box side panel with illustration of action. *Don Hulltzman Collection.*

# FLYING SKY PATROL

Battery operated RED FIRE CHIEF or GREEN POLICE HIGHWAY PATROL HELICOPTER, WITH MULTIPLE ACTIONS INCLUDING SIMULATED TAKE-OFF. Tin with plastic rotors and rubber wheels.
Toy Marks: T.P.S. (tail, right side), T.P.S. (molded on end of main blade).
Size: L: 13.75in W: 4.25in H: 5in (L: 35cm W: 11.5cm H: 13cm)
Introduced: c.1975. Est. quantity: 12,000 pieces each. Scarcity rating: 4.
*Fine to Excellent (C6 to C8): $150-$200, Mint (C10): $275.*

Box text: "FLYING SKY PATROL HELICOPTER," "DYNAMIC ACTION * AUTOMATIC ACTIONS * REVOLVING PROPELLER * TAKE-OFF & LANDING * GO-STOP * FLYING AROUND * REAL ENGINE NOISE." Box marks: T.P.S.

**Action:** Helicopter moves forward with revolving blades, blinking lights, and engine noise. Pedestal lowers from helicopter to raise it up from the ground in lift-off fashion. Helicopter spins as if flying then lowers to repeat actions.

# FORD MUSTANG

Friction 1970 RED 2 DOOR HARDTOP FORD MUSTANG WITH INTERIOR DETAIL AND OPEN SIDE WINDOWS. Tin with plastic wheels.
Toy Marks: T.P.S. (rear window deck).

**Action:** Friction action with inertia driven motor
Size: L: 10.5in H: 3.25in W: 4.25in (L: 27cm H: 8cm W: 11cm)
Introduced: c.1970. Est. quantity: 6,000 pieces. Scarcity rating: 4.5
*Fine to Excellent (C6 to C8): $150-$190, Mint (C10): $250.*

# FORD MUSTANG STUNT CAR

Battery operated ROLL OVER RED & BLACK 1970 FORD MUSTANG WITH "STUNT" AND "MACH 1" MARKINGS. Tin with plastic wheels.
Toy Marks: T.P.S. (rear window ledge).
Size: L: 10.75in H: 3.25in W: 4.25in (L: 27.5cm H: 8cm W: 11cm)
Introduced: c.1971. Est. quantity: 100,000 pieces. Scarcity rating: 2.
*Fine to Excellent (C6 to C8): $100-$135, Mint (C10): $175.*

Box text: "BATTERY OPERATED FORD MUSTANG STUNT CAR," "STEER IT YOURSELF AND PLAY." Box marks: T.P.S.

**Action:** Car goes forward then raises up on driver side wheels via cam and lever mechanism. Car barrel rolls 360° and goes forward again. Moveable front wheels for setting direction.

# GASOLINE CAR - MOBILGAS

Windup TRUCK CAB WITH ATTACHED MOBILGAS TANKER and fixed key. Tin with rubber tires. Known variations: Same cab with auto transport trailer, three separate cars on upper deck and lithographed images of cars on bottom of traile
Toy Marks: T.P.S. (right rear side of tanker)
Size: L: 10in H: 2.25in W: 1.5in (L: 25.5cm H: 6cm W: 4cm)
Introduced: c.1970. Est. quantity: 10,000 pieces. Scarcity rating: 4.
*Fine to Excellent (C6 to C8): $75-$95, Mint (C10): $125.*

Box text: "GASOLINE CAR WIND-UP TOY" (English and Japanese). Box marks: T.P.S.

**Action:** Truck cab has windup mechanism and moves forward with articulated tank trailer permanently attached.

Windup GAY 90'S STYLE CYCLIST ON HIGH WHEEL TRICYCLE, WITH BELL
MARKED 1890, with fixed key. Tin with felt jacket and satin like pants.
Toy Marks: T.P.S. (back of bell).
Size: H: 7in W: 2in L: 4.75in (H: 18cm W: 5cm L: 12cm)
Introduced: c.1955. Est. quantity: 100,000 pieces. Scarcity rating: 2.5.
*Fine to Excellent (C6 to C8): $225-$300, Mint (C10): $375.*

Box text: "GAY 90'S CYCLIST * MECHANICAL * WITH BELL." Box marks: T.P.S.

**Action:** Cycle with ringing bell moves in a wide circle. Jointed legs of the cyclist
move up and down in pedaling action.

Detail of bell lithography.

"Gay 90's Cyclist" Japanese Patent grant # 118245.
*Courtesy of Toplay (T.P.S.) Ltd.*

# GIRL SKIPPING ROPE

Windup GIRL AND BOY PLAYING SKIP ROPE WITH
SMALLER GIRL, with fixed key. Tin with vinyl heads on girl
Toy Marks: T.P.S. (base, behind boy).
Size: L: 12.5in H: 5.5in W: 2.25in (L: 31.5cm H: 14cm
W: 5.5cm)
Introduced: c.1960. Est. quantity: 50,000 pieces.
Scarcity rating: 3.
*Fine to Excellent (C6 to C8): $150-$200, Mint (C10): $275*

Box text: "MECHANICAL GIRL SKIPPING ROPE." Bo
marks: T.P.S. *Brynne & Scott Shaw Collection.*

**Action:** Girl and boy spin rope with their right hands
while smaller girl, attached by rod, jumps over the
rope. *Brynne & Scott Shaw Collection.*

Detail drawing of "Girl Skipping Rope."
Japanese patent applied for 1960, granted
1962. *Courtesy of Toplay (T.P.S.) Ltd.*

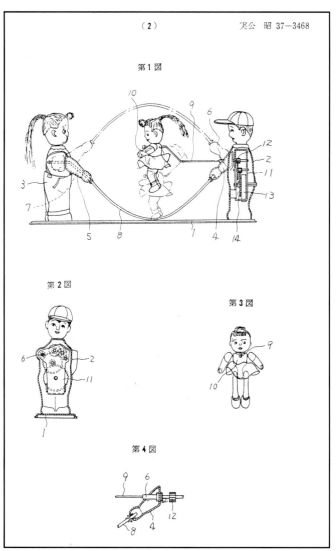

# GIRL WITH CHICKENS
## a.k.a. GIRL FEEDING CHICKENS

Windup GIRL FEEDING CHICKENS ATTACHED TO ROD, with fixed key. Tin with cloth head scarf.
Toy Marks: T.P.S. (below key).
Size: H: 5in W: 2.25in L: 5.5in (H: 13cm W: 6cm L: 14cm)
Introduced: c.1960. Est. quantity: 30,000 pieces. Scarcity rating: 3.
*Fine to Excellent (C6 to C8): $250-$300, Mint (C10): $400.*

Box text: "MECHANICAL GIRL WITH CHICKENS." Box marks: T.P.S., Rosko. Known box variation: "GIRL FEEDING CHICKENS," add 10%.
*Brynne and Scott Shaw Collection.*

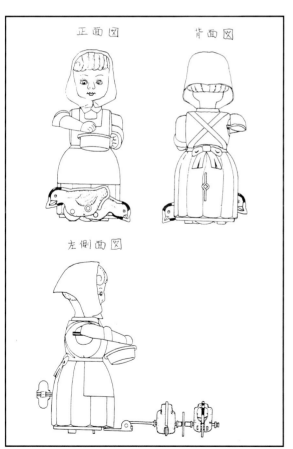

Multiple view outline drawing. *Courtesy of Toplay (T.P.S.) Ltd.*

**Action:** Girl walks sideways with swivel motion while chickens follow on rod. Right hand moves as if throwing seed, as her head bobs.

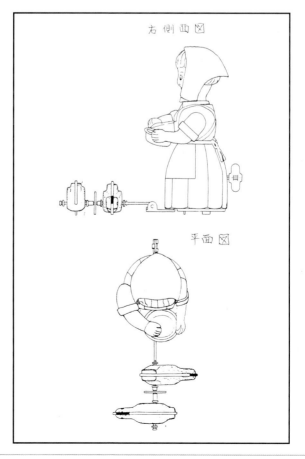

# GOOD COMPANION

KISSING COUPLE ON STONE WALL WITH DOG. Lever windup. Tin.
Size: L: 8in (L: 20.5cm)
Introduced: c.1965. Est. quantity: 6,000 pieces. Scarcity rating: 5.
*Fine to Excellent (C6 to C8): $200-$275, Mint (C10): $375.*

**Action:** Lever moves boy away from the girl and winds toy.
When released, he slowly moves toward the girl and kisses her
with kissing sound.

# GOOFY CYCLIST

Windup GOOFY ON HIGH WHEEL TRICYCLE WITH BELL SHOWING
IMAGES OF MINNIE AND PLUTO, with fixed key. Tin with satin like pants and
rubber ears.
Toy Marks: LINEMAR (bell), Walt Disney Productions (bell).
Size: H: 7in W: 2.5in L: 4.75in (H: 18cm W: 6.5cm L: 12cm)
Introduced: c.1958. Est. quantity: 10,000 pieces. Scarcity rating: 4.5.
*Fine to Excellent (C6 to C8): $1,800-$2,500, Mint (C10): $3,200.*

Box text: "MECHANICAL GOOFY CYCLIST * WALT DISNEY PRODUC-
TIONS * J-1733." Box marks: LINEMAR, Walt Disney Productions.

**Action:** Cycle with ringing bell moves in a wide circle. Goofy's
jointed legs move up and down in pedaling action.

Bell lithography with Minnie Mouse and Pluto.

# GOOFY THE UNICYCLIST

Windup GOOFY PEDALING ON UNICYCLE with fixed key. Tin with satin like pants and rubber ears.
Toy Marks: LINEMAR (paper label on bottom), Walt Disney Productions (paper label on bottom).
Size: H: 5.75in W: 2.25in D: 3.25in (H: 14.5cm W: 6cm D: 8cm)
Introduced: c.1958. Est. quantity: 20,000 pieces. Scarcity rating: 4.
*Fine to Excellent (C6 to C8): $950-$1,250, Mint (C10): $1,600.*

Box text: "MECHANICAL GOOFY THE UNICYCLIST." Box marks: LINEMAR, Walt Disney Productions. *Photo courtesy of New England Auction Gallery.*

**Action:** Goofy pedals forward on unicycle, spins around and moves forward again in different direction. Lever which extends down from the base lifts one wheel to make the toy turn.

## GRENDIZER STAND UP CYCLE

Battery operated JAPANESE CHARACTER UFO ROBO GRENDIZER© STANDING ON 3-WHEEL MOTORCYCLE. UFO Robo Grendizer is the third character in the Majinga series. Plastic with tin overlay on top of cycle.
Toy Marks: T.P.S. (molded in front end of base).
Size: L: 10in H: 5.5in W: 5in (L: 25.5cm H: 14cm W: 13cm)
Introduced: c.1975. Est. quantity: 20,000 pieces. Scarcity rating: 4.
*Fine to Excellent (C6 to C8): $125-$175, Mint (C10): $225.*

Box text: "BATTERY OPERATED UFO ROBO GRENDIZER STAND UP CYCLE * DYNAMIC ACTION" (In Japanese).

**Action:** Grendizer is positioned on cycle which travels forward in direction set by front wheel. Cycle raises up on back wheels via drop down lever with wheel, then spins in left hand circle before lowering and repeating actions.

## GRUMMAN F-14A JET FIGHTER TOMCAT

Box text: "GRUMMAN F-14A JET FIGHTER 'TOM-CAT' WITH AUTOMATIC * STOP AND GO ACTION * JET ENGINE NOISE * CABIN LIGHT AND ENGINE LIGHTS * VARIABLE SWEEP." Box marks: MESEM.

**Action:** F14A, with variable sweeping wings, taxis rapidly in circle with flashing engine exhaust, cockpit lights, and engine sound. Plane then stops before repeating cycle.

Battery operated F14A TOMCAT JET WITH MOVEABLE WINGS. Plastic.
Toy Marks: T.P.S. (molded on battery cover), MESEM (molded on battery cover).
Size: L: 10.25in W: 6.75in H: 3.5in (L: 26cm W: 17cm H: 9cm)
Introduced: c.1979. Est. quantity: 60,000 pieces. Scarcity rating: 2.5.
*Fine to Excellent (C6 to C8): $45-$55, Mint (C10): $75.*

# HAPPY HIPPO

Windup NATIVE WITH BANANAS RIDING ON HIPPOPOTAMUS, with fixed key.
Tin. Known variations: Gray colored hippo, add 15-20%.
Toy Marks: T.P.S. (bottom of jaw).
Size: L: 6in H: 4in W: 2.25in (L: 15cm H: 10cm W: 5.5cm)
Introduced: c.1957. Est. quantity: 60,000 pieces. Scarcity rating: 3.5.
*Fine to Excellent (C6 to C8): $350-$450, Mint (C10): $575.*

Box text: "MECHANICAL HAPPY HIPPO." Box marks: T.P.S. *Don Hultzman Collection.*

**Action:** Hippo goes in circles while opening its mouth trying to reach bananas on a pole held by native. Bump and go action with start/stop lever. *Don Hultzman Collection.*

Gray colored hippo. *Photo courtesy of Pete Thompson.*

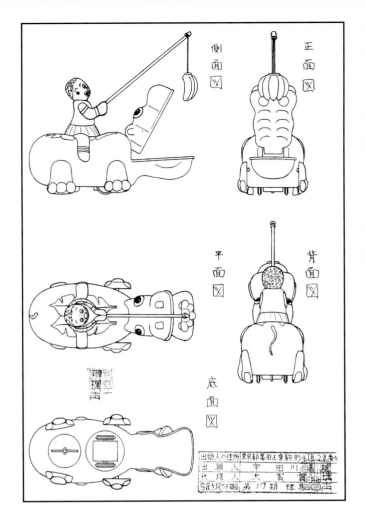

"Happy Hippo." Japanese Patent # 144020. *Courtesy of Toplay (T.P.S.) Ltd.*

## HAPPY PLANE

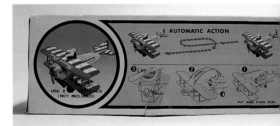

Battery operated CHAMPION BI-PLANE WITH REMOVABLE WINGS, MARKED 'HAPPY PLANE,' TAXIS IN TURNING PATTERN. Tin with plastic fuselage and pilot.
Toy Marks: T.P.S. (rudder, left side).
Size: L: 9.5in H: 3.75in W: 10.5in (L: 24cm H: 9.5cm W: 26.5cm)
Introduced: c.1975. Est. quantity: 20,000 pieces. Scarcity rating: 3.
*Fine to Excellent (C6 to C8): $100-$130, Mint (C10): $170.*

Box text: "BATTERY OPERATED AUTOMATIC ACTION HAPPY PLANE." Box marks: T.P.S.

**Action:** Biplane, with spinning propeller, taxis forward th loops around in circle before taxiing forward again.

Windup ROLLER SKATING BEAR with fixed key. Tin with satin like pants and rubber ears. Known variations: rigid felt ears.
Toy Marks: T.P.S. (back left shoulder).
Size: H: 6.25in W: 3.25in D: 4in (H: 16cm W: 8cm D: 10cm)
Introduced: c.1958. Est. quantity: 30,000 pieces. Scarcity rating: 3.5.
*Fine to Excellent (C6 to C8): $225-$300, Mint (C10): $375.*

Box text: "MECHANICAL HAPPY SKATERS" (with BEAR picture). Box marks: T.P.S., Cragstan. *Brynne and Scott Shaw Collection.*

**Action:** Bear skates on wheeled left foot with right leg pushing off, causing body to dip in realistic skating motion. *Brynne and Scott Shaw Collection.*

## HAPPY SKATERS - RABBIT

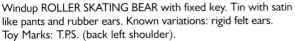

Windup ROLLER SKATING RABBIT with fixed key. Version 1: Tin with satin like pants and rubber ears. Known variations: rigid felt ears.
Toy Marks: T.P.S. (left side of back).
Size: H: 6.25in W: 3.25in D: 4in (H: 16cm W: 8cm D: 10cm)
Introduced: c.1958. Est. quantity: 20,000 pieces (total of version 1 and 2). Scarcity rating: 3.5.
*Fine to Excellent (C6 to C8): $250-$325, Mint (C10): $425.*

Version 2: Tin with plaid pants, rubber ears, felt jacket and bow tie.
Toy Marks: Made in Japan. (outside of right skate).
Introduced: c.1960. Add 10-15%

**Action:** Rabbit skates on wheeled left foot with right leg pushing off, causing body to dip in realistic skating motion.

Box text: "MECHANICAL HAPPY SKATERS" (with RABBIT picture).
Box marks: T.P.S., Cragstan. *Photo courtesy of Pete Thompson.*

# HAPPY, THE VIOLINIST (Version 1)

STAR EYED VIOLIN PLAYING CLOWN ON STILTS WITH BOW TIE, RED JACKET, AND SHOES. Windup with fixed key. Tin with felt jacket, cloth pants and bow tie.
Toy Marks: None.
Size: H: 9.25in W: 2.25in D: 2.75in (H: 23.5 cm W: 6cm D: 7cm)
Introduced: c.1955. Est. quantity: 12,000 pieces. Scarcity rating: 4.
*Fine to Excellent (C6 to C8): $250-$325, Mint (C10): $450.*

Box text: "MECHANICAL HAPPY, THE VIOLINIST." *Photo courtesy of New England Auction Gallery.*
**Action:** Violinist sways from side to side while right arm moves bow across violin. Long legs are stabilized by big feet.

# HAPPY, THE VIOLINIST (Version 2)

VIOLIN PLAYING CLOWN ON STILTS WITH GREEN JACKET, STRIPED PANTS, AND BLACK SHOES. Windup with fixed key. Tin with satin like pants and felt jacket.
Toy Marks: T.P.S. (top of violin near face).
Size: H: 8.75in W: 2.5in D: 2.5in (H: 22.5cm W: 6.5cm D: 6.5cm)
Introduced: c.1956. Est. quantity: 100,000 pieces. Scarcity rating: 2.
*Fine to Excellent (C6 to C8): $150-$200, Mint (C10): $275.*

Box text: "Mechanical HAPPY THE VIOLINIST," "IT'S CRAGSTAN FOR TOYS 1216." Box marks: T.P.S., Cragstan, HTC. Box also seen without Cragstan.

**Action:** Violinist sways from side to side while right arm moves bow across violin. Long legs are stabilized by big feet.

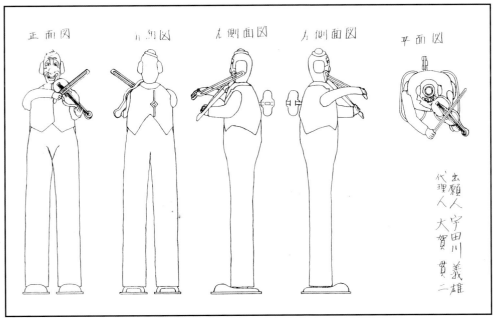

Front, back, side, and top views. *Courtesy of Toplay (T.P.S.) Ltd.*

# HELICOPTER ON AIRFIELD

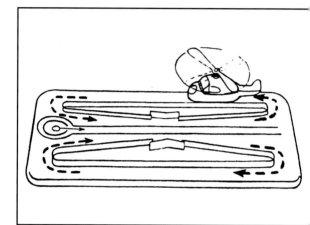

Windup 9cm (3.5in) HELICOPTER ON PLATFORM BASE WITH AIRFIELD AND COUNTRYSIDE SCENES, with separate key. Tin with plastic helicopter blade.
Toy Marks: T.P.S. (left side of base), Hirata (bottom right side of base).
Size: 9.25in x 5.5in platform (23.5cm x 14cm platform)
Introduced: c.1965. Est. quantity: 30,000 pieces. Scarcity rating: 3.
*Fine to Excellent (C6 to C8): $100-$125, Mint (C10): $150.*

Box text: "CRAGSTAN HELICOPTER ON AIRFIELD," "A WIND-UP HELICOPTER WITH SPINNING PROPELLER * LOTS OF SPEEDY ACTION ON THE AIRFIELD." Box marks Cragstan, T.P.S., NGS. Box also seen without T.P.S. logo.

**Action:** Helicopter with spinning propeller goes around airfield platform hooked on guide rail in figure eight pattern.

# HELICOPTER WITH AUTOMOBILE

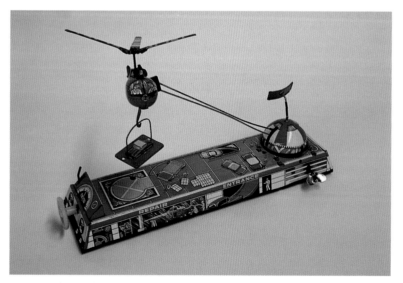

FLYING HELICOPTER CARRYING CAR ABOVE BASE WITH CONTROLS. Windup with fixed key. Tin with plastic control wheel.
Toy Marks: T.P.S. (left side below radar tower).
Size: L: 12.25in W: 3.25in H: 7.5in (L: 31cm W: 8cm H: 19cm)
Introduced: c.1961. Est. quantity: 6,000 pieces. Scarcity rating: 4.5.
*Fine to Excellent (C6 to C8): $500-$650, Mint (C10): $850.*

**Action:** Helicopter flies in circular pattern around base, attached by double rod. Altitude is controlled by front steering wheel to allow helicopter to hook and carry a platform loaded with a car. Separate key and windup motor controls helicopter blades. Start-stop lever.

# HIGHTECHNICAL BIG RIDER

Battery operated GREEN HONDA MOTORCYCLE #34 AND RIDER, WITH SPIN OUT ACTION. Plastic with tin body. Plastic tires and rubber drive wheels. Known variations: See "Hightechnical Rider."
Toy Marks: T.P.S. (top, below gauges).
Size: L: 10in H: 6.25in W: 4in (L: 25.5cm H: 16cm W: 10cm)
Introduced: c.1977. Est. quantity: 15,000 pieces. Scarcity rating: 4.
*Fine to Excellent (C6 to C8): $135-$175, Mint (C10): $225.*

Box text: "BATTERY OPERATED HIGHTECHNICAL BIG RIDER * DYNAMIC ACTION!!." Box marks: T.P.S.

**Action:** Motorcycle goes forward in circular pattern then spins out during banking turn as lever extends from bottom. Motorcycle then uprights itself via lever arm on the side and repeats actions.

# HIGHTECHNICAL RALLY

Battery operated PORSCHE 911-S RALLY CAR WITH RACING LITHO #17. Tin with rubber wheels.
Toy Marks: T.P.S. (dashboard, right side).
Size: L: 10in H: 3.25in W: 4in (L: 25cm H: 8cm W: 10cm)
Introduced: c.1971. Est. quantity: 20,000 pieces. Scarcity rating: 4.
*Fine to Excellent (C6 to C8): $125-$160, Mint (C10): $200.*

Box text: "BATTERY OPERATED HIGHTECHNICAL RALLY * HIGH-TECHNICAL AUTOMATIC ACTION." Box marks: T.P.S.

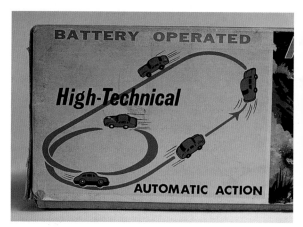

**Action:** Car travels forward a short distance then raises up via cam and lever and makes an approximate 180° turn on two inside wheels. Car lowers and travels forward as action repeats but with a 360° turn.

Action illustration on box side panel.

## HIGHTECHNICAL RIDER

Battery operated HONDA RACING MOTORCYCLE #33 AND RIDER WITH SPIN OUT ACTION. Plastic with tin body and rubber tires and drive wheels. Known variations: Rider with blue or orange and brown outfit as shown above. Toy Marks: T.P.S. (left side near on-off switch).
Size: L: 10in H: 6.25in W: 4in (L: 25.5cm H: 16cm W: 10cm)
Introduced: c.1973. Est. quantity: 30,000 pieces. Scarcity rating: 3.5.
*Fine to Excellent (C6 to C8): $135-$175, Mint (C10): $225.*

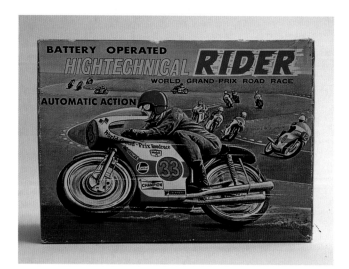

**Action:** Motorcycle goes forward in circular pattern then spins out during banking turn as lever extends from bottom. Motorcycle then uprights itself via lever arm on the side and repeats actions.

Box text: "BATTERY OPERATED HIGHTECHNICAL RIDER," "WORLD GRAND PRIX ROAD RACE * AUTOMATIC ACTION." Box marks: T.P.S.

# HIGHWAY PATROL CAR

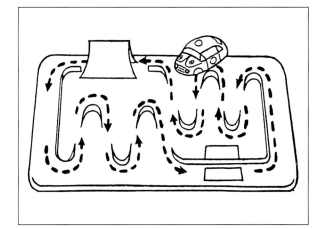

Windup 6cm (2.25in) PATROL CAR ON PLATFORM BASE WITH POLICE AND GANGSTER SCENES AND TUNNEL, with separate key. Tin.
Toy Marks: T.P.S. (left lower side of platform), Hirata (left lower side of platform).
Size: 15.25in x 6.75in platform (38.5cm x 17cm platform)
Introduced: c.1965. Est. quantity: 6,000 pieces. Scarcity rating: 4.5.
Fine to Excellent (C6 to C8): $175-$225, Mint (C10): $300.

Box text: "MECHANICAL HIGHWAY PATROL CAR." Box marks: T.P.S.

**Action:** Windup Highway Patrol car goes around platform and through tunnel hooked on zigzag guide rail.

# HIGHWAY SET

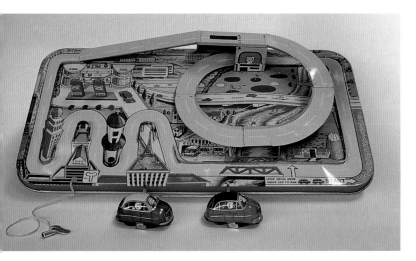

TWO 6cm (2.25in) WINDUP AUTOMOBILES ON PLATFORM BASE WITH ELEVATED ROAD AND CITY AND COUNTRY SCENES, INCLUDING ESSO GAS PUMPS AND REFINERY. Separate key. Tin with rubber and plastic wheels on cars. Known variations: LUCKY HIGHWAY SET with base mounted key.
Toy Marks: T.P.S. (upper right side of base), Hirata (upper right side of base).
Size: L: 16.75in W: 8.75in H: 2.25in (L: 42.5cm W: 22cm H: 6cm)
Introduced: c.1965. Est. quantity: 10,000 pieces. Scarcity rating: 4.
Fine to Excellent (C6 to C8): $150-$200, Mint (C10): $250.

Box text: "MECHANICAL HIGHWAY SET WITH 2 CARS * AUTOMATIC INTERCHANGE *N0.5763." Box marks: T.P.S.

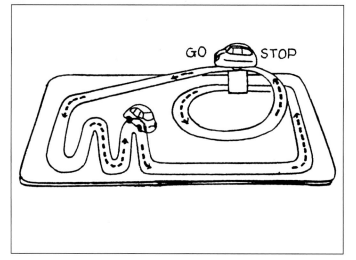

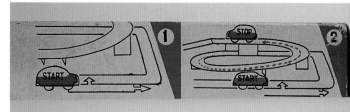

Box illustration of action.

**Action:** First car travels around track hooked on rail and stops on the elevated highway. Second car, when placed on track, travels to point below first car and triggers a release of the stopped car. This allows action to continue with proper spacing between cars.

## HIKARI EXPRESS RAIL BOARD

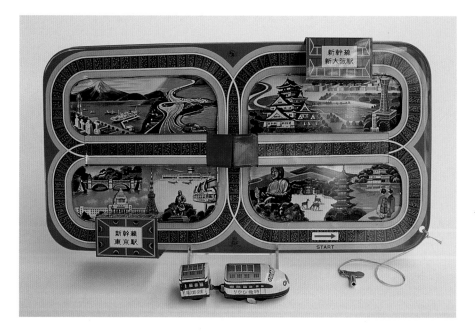

Windup 2 PIECE 10.5cm (4.25in) BULLET TRAIN (SHINKANSEN) ON OVER/UNDER FIGURE EIGHT PLATFORM BASE, WITH JAPAN SCENERY AND TOKYO AND OSAKA STATION TUNNELS. Separate key. Tin with rubber and plastic wheels.
Toy Marks: T.P.S. (bottom center of base), Hirata (top center of base).
Size: L: 16.75in W: 8.75in H: 2.25in (L: 42.5cm W: 22cm H: 5.5cm)
Introduced: c.1970. Est. quantity: 24,000 pieces.
Scarcity rating: 4.
*Fine to Excellent (C6 to C8): $150-$200, Mint (C10): $250.*

Box text: "HIKARI EXPRESS BOARD RAIL," "TOKYO TO SHIN OSAKA 3 HOURS AND 10 MINUTES * LINE WITH VIEWS." Box marks: T.P.S.

**Action:** Windup train goes around platform and through station tunnels hooked on rail. Track elevates and crosses over itself via opening bridge, in figure eight fashion.

# HUNGRY CAT

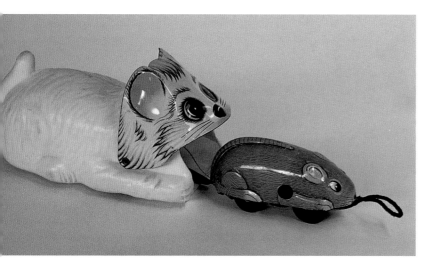

Windup CAT CHASING AND SWALLOWING MOUSE. Plastic cat with tin mouth and tin mouse.
Toy Marks: T.P.S. (right rear side of mouse), Japan (left side of cat's mouth).
**Action:** As mouse is pulled out of cat's mouth, it winds the mechanism with attached nylon line. As mouse moves forward, the line is pulled in bringing the cat closer and giving the appearance of swallowing the mouse.
Size: L: 5.5in H: 2.25in W: 2.25in (L: 14cm H: 6cm W: 6cm)
Introduced: c.1970. Est. quantity: 12,000 pieces.
Scarcity rating: 4.
*Fine to Excellent (C6 to C8): $75-$95, Mint (C10): $125.*

# HUNGRY WHALE

Windup WHALE FOLLOWING AND SWALLOWING SMALLER FISH. Plastic whale with tin mouth and tin fish.
Toy Marks: Japan (tail of fish and mouth of whale).
Size: L: 6.25in H: 2.25in W: 2.25in (L: 16cm H: 6cm W: 5.5cm)
Introduced: c.1965. Est. quantity: 30,000 pieces. Scarcity rating: 3.
*Fine to Excellent (C6 to C8): $65-$85, Mint (C10): $100.*

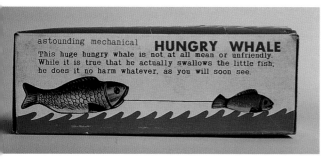

Box text: "HUNGRY WHALE," "astounding mechanical * swallows a whole fish at one gulp in front of your unbelieving eyes." Box marks: T.P.S. or Shackman.

Box variations: T.P.S., the manufacturer, or Shackman, the importer.

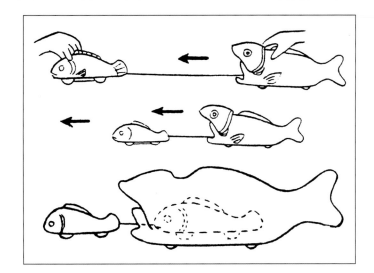

**Action:** As fish is pulled out of whale's mouth, it winds the mechanism with attached nylon line. As fish moves forward, the line is pulled in, bringing the whale closer and giving the appearance of swallowing the fish.

# JOE THE ACROBAT

Windup HOBO CLOWN BALANCING ON HIS HANDS ON TOP OF MOVING BALL, with separate key. Tin with satin like pants.
Toy Marks: T.P.S. (back side of ball).
Size: H: 6.75in W: 2.25in D: 3.5in (H: 17cm W: 6cm D: 9cm)
Introduced: c.1956. Est. quantity: 12,000 pieces. Scarcity rating: 4.5.
*Fine to Excellent (C6 to C8): $475-$600, Mint (C10): $750.*

Box text: "MECHANICAL JOE THE ACROBAT," "PAT.PEND.." Box marks: T.P.S., HTC.

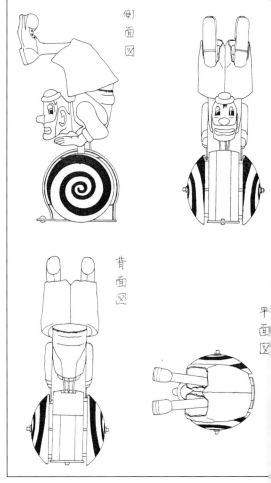

**Action:** Clown hand stands on ball with spinning sides. Hands move like they are spinning the ball and jointed legs move up and down.

Side, front rear, and top views. Japanese Patent # 128038, granted 1957. *Courtesy of Toplay (T.P.S.) Ltd.*

# JOE THE XYLOPHONE PLAYER

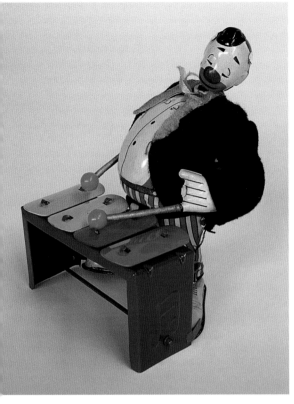

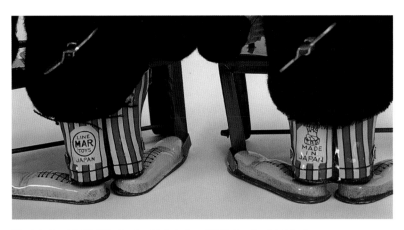

Toy Marks: LINEMAR (back of left leg) or T.P.S. (back of left leg).

Box text: "MECHANICAL JOE THE XYLOPHONE PLAYER." Box marks: LINEMAR or T.P.S. *Brynne and Scott Shaw Collection.*

**Action:** Clown with jointed head turns from side to side and plays across xylophone with arms moving up and down. Sticks are springs connected to hands.

Windup CLOWN WITH BIG FEET PLAYING 4 NOTE XYLOPHONE, with fixed key. Tin with felt jacket and spring sticks. Known variations: First version has small hands (15mm) with stick extending from left hand index finger and right hand fourth finger. Second version has large hands (19mm) with sticks extending from little finger on both hands.
Size: H: 5in W: 3.5in D: 3.5in (H: 13cm W: 9cm D: 9cm)
Introduced: c.1958. Est. quantity: 50,000 pieces. Scarcity rating: 3.
*Fine to Excellent (C6 to C8): $300-$375, Mint (C10): $450.*

# JOLLY LOCO

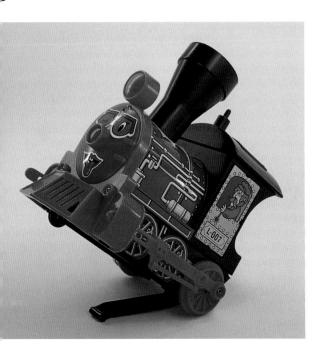

Battery operated COMIC LOCOMOTIVE WITH BOUNCING SMOKESTACK AND FORWARD AND BACK STANDUP ACTION.
Plastic with tin boiler and side panels.
Toy Marks: T.P.S. (right side of tin boiler), T.P.S. (molded on bottom).
Size: L: 6.25in H: 5.5in W: 4in (L: 16cm H: 14cm W: 10cm)
Introduced: c.1977. Est. quantity: 20,000 pieces. Scarcity rating: 3.5.
*Fine to Excellent (C6 to C8): $45-$60, Mint (C10): $75.*

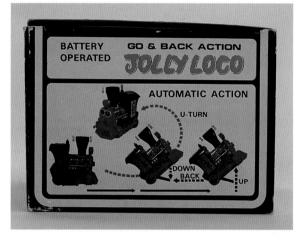

Box text: "BATTERY OPERATED GO & BACK ACTION JOLLY LOCO," "AUTOMATIC ACTION." Box marks: T.P.S.

**Action:** Loco goes forward in direction set by moveable guide wheel, stops and raises up on back wheels via lever, backs up and lowers before repeating actions. Bouncing smokestack.

# JOLLY PLANE

Battery operated BUMP-N-GO, YELLOW AND BLUE PLANE WITH FOLDING WINGS AND PILOT. Tin and plastic body, pilot, and propellers.
Toy Marks: T.P.S. (left side of fuselage above wing).
Size: L: 6.75in W: 4.75in H: 3.25in (L: 17cm W: 12cm H: 8cm)
Introduced: c.1977. Est. quantity: 30,000 pieces. Scarcity rating: 3.
*Fine to Excellent (C6 to C8): $125-$150, Mint (C10): $185.*

Box text: "BATTERY OPERATED JOLLY PLANE." Box marks: T.P.S.

**Action:** Bump and go action while propeller spins and pistons move. Has on-off switch.

# JOLLY WIGGLING SNAKE

Windup COILED SNAKE WITH WIGGLING HEAD AND TAIL, with fixed key. Tin with rubber wheels.
Toy Marks: T.P.S. (on base, right side of tail).
Size: H: 5.5in W: 3.75in L: 7.5in (H: 14cm W: 9.5cm L: 19cm)
Introduced: c.1960. Est. quantity: 12,000 pieces. Scarcity rating: 4.
*Fine to Excellent (C6 to C8): $130-$160, Mint (C10): $200.*

Box text: "MECHANICAL JOLLY WIGGLING SNAKE," "WIGGLING ACTION * HEAD TURNS FROM SIDE TO SIDE." Box marks: T.P.S.

**Action:** Head turns left and right and tail moves back and forth as snake travels in wide circle. Head detaches from body to fit in box.

# JUGGLING CLOWN (WITH APPLES)

Windup BIG FOOTED CLOWN JUGGLING PLATE OF APPLES ON NOSE WITH HAT IN RIGHT HAND, with fixed key. Tin with felt jacket. Known variations: See "Juggling Clown" (with ball).
Toy Marks: T.P.S. (side of left leg) or LINEMAR.
Size: H: 8.75in W: 4in D: 3.5in (H: 22cm W: 10cm D: 9cm)
Introduced: c.1956. Est. quantity: 15,000 pieces. Scarcity rating: 4.
*Fine to Excellent (C6 to C8): $400-$500, Mint (C10): $650.*

Box text: "MECHANICAL JUGGLING CLOWN." Box marks: T.P.S. or LINEMAR.

**Action:** Clown rotates and rocks back and forth with arms spinning rapidly, as he tries to balance plate of apples on nose. Pole is detachable and connected to nose by a spring.

# JUGGLING CLOWN (WITH BALL)

Windup BIG FOOTED CLOWN JUGGLING BALL ON NOSE WITH HAT IN RIGHT HAND, with fixed key. Tin with felt jacket. Known variations: See "Juggling Clown" (with apples).
Toy Marks: T.P.S. (side of left leg) or LINEMAR.
Size: H: 8.75in W: 4in D: 3.5in (H: 22cm W: 10cm D: 9cm)
Introduced: c.1956. Est. quantity: 15,000 pieces. Scarcity rating: 4.
*Fine to Excellent (C6 to C8): $400-$500, Mint (C10): $625.*

Box text: "MECHANICAL JUGGLING CLOWN." Box marks: T.P.S. or LINEMAR.

**Action:** Clown rotates and rocks back and forth with arms spinning rapidly, as he tries to balance ball on his nose. Pole is detachable and connected to nose by a spring.

"Juggling Clowns" side by side.

# JUGGLING DUCK & HIS FRIENDS

Windup CIRCUS PARADE DUCK WITH TIN SPINNERS PULLING LEAF WAGON WITH 3 SQUIRRELS, with fixed key. Tin. Known variations: Spinners made of plastic (subtract15%). Same toy as "Slim the Seal & his Friends," but with different lithography.
Toy Marks: T.P.S. (rear of leaf wagon).
Size: L: 9.75 H: 4.75in W: 4.25in (L: 25cm H: 12cm W: 11cm)
Introduced: c.1960. Est. quantity: 12,000 pieces. Scarcity rating: 4.
*Fine to Excellent (C6 to C8): $300-$375, Mint (C10): $500.*

Box text: "CIRCUS PARADE JUGGLING DUCK & his FRIENDS * MECHANICAL ACTION TOY * HE JUGGLES & HUSTLES." Box marks: T.P.S. *Photo courtesy of Herb Smith, Smith House Toys.*
**Action:** Duck with circus litho twirls spinners while moving forward and pulling wagon with one stationary and two spinning squirrels.

# JUGGLING POPEYE AND OLIVE OYL

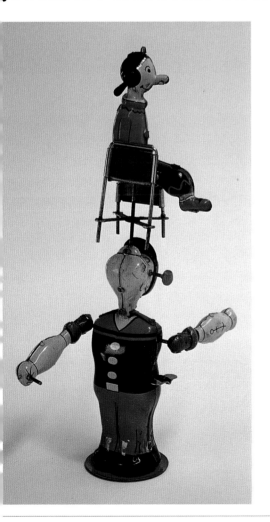

Windup POPEYE TWIRLING OLIVE OYL ON CHAIR, with fixed key. Tin with rubber hair bun on Olive Oyl. Based on "Clown Juggler with Monkey" design.
Toy Marks: LINEMAR (back of Popeye), King Features Synd. (back of Popeye).
Size: H: 9in W: 3.5in D: 2.25in (H: 23cm W: 9cm D: 6cm)
Introduced: c.1960. Est. quantity: 20,000 pieces. Scarcity rating: 4.
*Fine to Excellent (C6 to C8): $2,400-$3,000, Mint (C10): $3,800.*

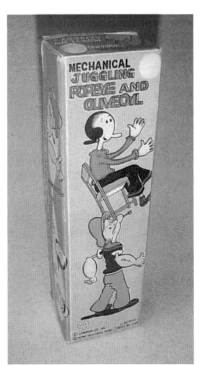

Box text: "MECHANICAL JUGGLING POPEYE AND OLIVEOYL." Box marks: LINEMAR. *Photo courtesy of Randy Ibey, Randy's Toy Shop.*

**Action:** Popeye's arms rotate. Right arm has stick which hits chair leg, causing Olive Oyl to spin while balanced on rod attached to his nose.

# JUMPING CAT

Windup CAT WITH IRREGULAR CAM LIKE WHEELS TO CREATE JUMPING EFFECT, with fixed key. Tin. *Japan Toys Museum Foundation.*
Toy Marks: T.P.S. (bottom, towards front).
**Action:** Serrated off-center wheels cause cat to move forward 5 jumps. Cat stops and turns head, left-right-left-right, then repeats actions.
Size: L: 5in H: 2.75in W: 2in (L: 13cm H: 7cm W: 5cm)
Introduced: c.1958. Est. quantity: 6,000 pieces. Scarcity rating: 4.5.
*Fine to Excellent (C6 to C8): $125-$160, Mint (C10): $200.*

# JUMPING SQUIRREL

Windup SQUIRREL WITH IRREGULAR CAM LIKE WHEELS TO CREATE JUMPING EFFECT, with fixed key. Tin with rubber ears.
Toy Marks: T.P.S. (bottom, towards front).
Size: L: 5.75in H: 4in W: 2.25in (L: 14.5cm H: 10cm W: 5.5cm)
Introduced: c.1957. Est. quantity: 12,000 pieces. Scarcity rating: 4.
*Fine to Excellent (C6 to C8): $125-$150, Mint (C10): $175.*

Box text: "MECHANICAL JUMPING SQUIRREL." Box marks: T.P.S.

**Action:** Serrated off-center wheels cause squirrel to move forward 5 jumps. Squirrel stops and turns head, left-right-left-right, then repeats actions.

# KINDERGARTEN BUS

Box text: "KINDERGARTEN BUS" (in Japanese and English). Box marks: T.P.S.

**Action:** Bus goes forward in straight direction with friction inertia movement.

Friction JAPANESE KINDERGARTEN BUS WITH ANIMAL DRAWINGS. Tin with plastic tires. Known variations: First version with interior seats, add 15%. Third version without children pictured in windows, subtract 15%.
Toy Marks: T.P.S. (right rear side panel).
Size: L: 10in H: 4.25in W: 4in (L: 25.5cm H: 11cm W: 10cm)
Introduced: c.1970. Est. quantity: 30,000 pieces. Scarcity rating: 3.
Fine to Excellent (C6 to C8): $65-$75, Mint (C10): $100.

# LADY BUG & TORTOISE WITH BABIES

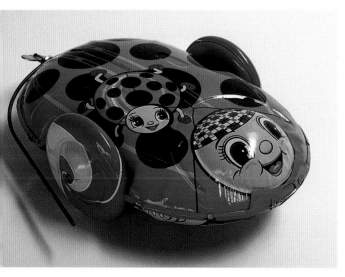

Ladybug side. FLIP-OVER TOY WITH LADY BUGS ON ONE SIDE AND TURTLES ON OTHER. Windup with fixed key. Tin.
Toy Marks: T.P.S. (rear, on Lady Bug side, near key)
Size: L: 6in W: 4in H: 2in (L: 15cm W: 10cm H: 5cm)
Introduced: c.1960. Est. quantity: 20,000 pieces. Scarcity rating: 3.
Fine to Excellent (C6 to C8): $75-$100, Mint (C10): $125.

Tortise side. **Action:** As toy moves forward on its wheels, a rod at the rear of the toy causes it to flip over showing either a ladybug or a turtle with baby.

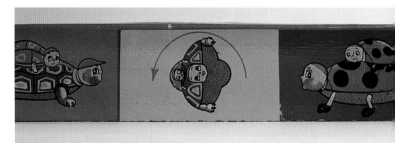

Box text: "MECHANICAL LADY BUG & TORTOISE WITH BABIES," "AS THE TOY MOVES, IT FLIPS OVER AUTO-MATICALLY TO BECOME EITHER A LADY BUG OR TORTOISE." Box marks: T.P.S.

Illustration of flip-over action.

## LADY-BUG FAMILY PARADE a.k.a. LADY-BUGS

Original version. MAMA LADY-BUG WITH THREE BABY ONES FOLLOWING IN A SINGLE LINE. Windup with fixed key. Tin with spring antennae and rubber wheels.
Toy Marks: T.P.S. (rear side of mama lady-bug)
Size: L: 13.5in W: 3.25in H: 1.75in (L: 34cm W: 8cm H: 4.5cm)
Introduced: c.1959. Est. quantity: 5,000,000 pieces. Scarcity rating: 1.
*Fine to Excellent (C6 to C8): $60-$75, Mint (C10): $100.*

Contemporary version. Plastic base, wheels and joining rods and lithographed (no spring) antennae. Many variations exist between the original 1959 version (600,000 pieces) and the contemporary version (subtract 30%). In addition, Korean and Chinese copies of this toy exist.

Box variations: "MECHANICAL LADY-BUG FAMILY PARADE," marked Cragstan, T.P.S. and "MECHANICAL LADY-BUG FAMILY PARADE" (in Japanese), marked T.P.S.

Box variations: "MECHANICAL LADY-BUGS," marked Rosko and "CRAGSTAN LADY-BUG FAMILY PARADE," marked Cragstan, T.P.S., NGS.

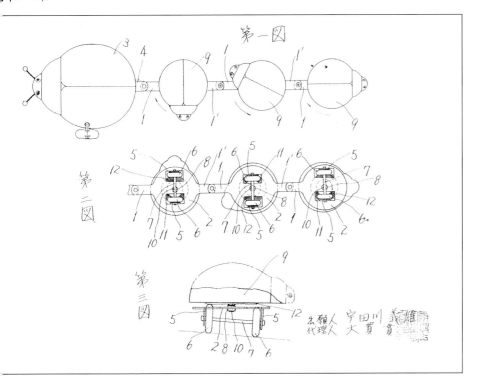

*Courtesy of Toplay (T.P.S.) Ltd.*
**Action:** Guide wheel causes lady-bug to move left and right alternately. The attached smaller lady-bugs follow and spin around individually via friction contact with wheels.

# LOCO CAB OVERLAND TRAFFIC GAME

Manual version of LOCOMOTIVE ON TRACK AND CAR ON HIGHWAY, CROSSING EACH OTHER ON PLATFORM BASE WITH CITY, COUNTRY, RIVER, CAMPING, SAFARI, AND INDIAN SCENES. Windup with separate key. Tin with rubber tires on vehicles. Known variations: Both a taxi and a police car have been used with this toy. Manual and automatic version.
Toy Marks: T.P.S. (lower left corner of base), Hirata (lower left corner of base).
Size: L: 16.5in W: 8.75in H: 2in (L: 42.5cm W: 22cm H: 5cm)
Introduced: c.1965. Est. quantity: 12,000 pieces.
Scarcity rating: 4.
*Fine to Excellent (C6 to C8): $125-$160, Mint (C10): $200.*

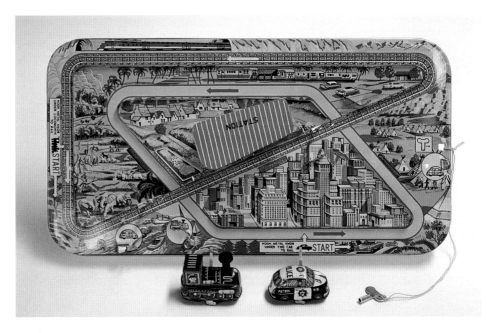

**Action:** Loco and car are equipped with a gravity, gear stopping mechanism which stops them when passing over slots in track. When the other vehicle passes over protruding tabs, which are attached to levers that fill the slots elsewhere on the base, this allows the stopped vehicle to proceed without a collision (automatic version). Crossing gates are spring loaded. Manual version controls action by pressing tabs to allow vehicles to proceed.

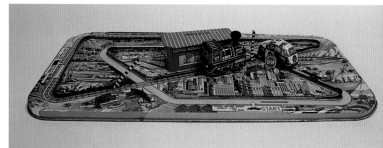

Automatic version.

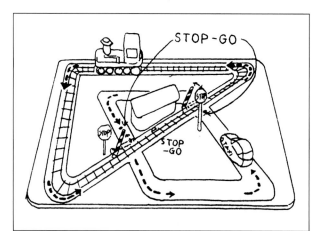

# LOOP THE LOOP COASTER

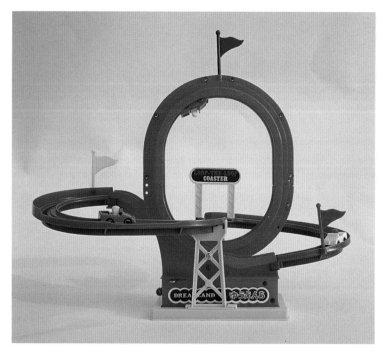

THREE 4cm (1.5in) CARS RACING AROUND ON LOOPING ROLLER COASTER. Battery operated. Plastic with brass wheels on cars. Known variations: Different coaster configurations.
Toy Marks: T.P.S. (Molded on battery cover).
Size: H: 13.5in W: 15.75in D: 9.5in (H: 34cm W: 40cm D: 24cm)
Introduced: c.1980. Est. quantity: 30,000 pieces. Scarcity rating: 3.
*Fine to Excellent (C6 to C8): $40-$55, Mint (C10): $75.*

Box text: "BATTERY OPERATED LOOP-THE-LOOP COASTER" (in Japanese and English). Box marks: T.P.S.

**Action:** A motor driven spring rotates in track and catches rod protruding from beneath cars to lift them up the inside of the loop. Cars are released to roll free around the track to the bottom and to repeat the action.

# LUCKY MONKEY PLAYING BILLIARDS

Windup POOL SHOOTING MONKEY WITH 3 POOL BALLS, with fixed key. Tin with one white and two red plastic pool balls.
Toy Marks: T.P.S. (left pant leg).
Size: L: 6.25in H: 4in W: 4.75in (L: 16cm H: 10cm W: 12cm)
Introduced: c.1960. Est. quantity: 20,000 pieces. Scarcity rating: 3.5.
Fine to Excellent (C6 to C8): $225-$275, Mint (C10): $350.

Box text: "MECHANICAL LUCKY MONKEY PLAYING BILLIARDS * ROSKO 0210." Box marks: T.P.S., Rosko

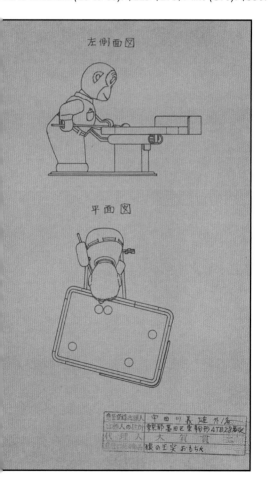

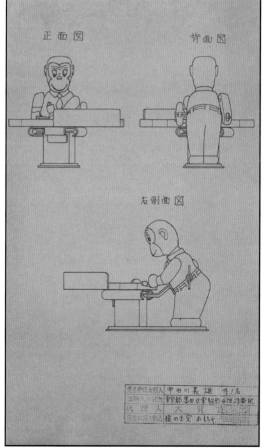

*Courtesy of Toplay (T.P.S.) Ltd.*

**Action:** Monkey lowers head, readies cue stick twice, and shoots pool ball into corner pocket as head snaps back in realistic action.

# LUNA HOVERCRAFT

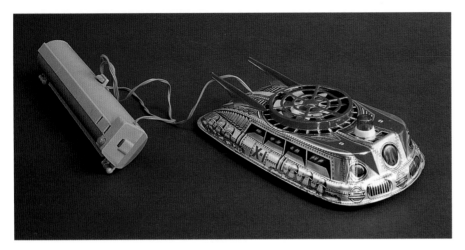

Battery operated RED, BLUE, AND SILVER FLOATING SPACESHIP WITH REMOTE BATTERY CONTROL. Aluminum with plastic light dome and plastic battery holder.
Toy Marks: T.P.S. (rear center of hovercraft).
Size: L: 7.75in W: 4.75in H: 2.25in (L: 20cm W: 12cm H: 5.5cm)
Introduced: c.1969. Est. quantity: 10,000 pieces. Scarcity rating: 4.
*Fine to Excellent (C6 to C8): $150-$200, Mint (C10): $275.*

Box text: "BATTERY OPERATED LUNA HOVERCRAFT," "NO WHEELS * DRIVING FORCE BY STRONG EXHAUSTION * RISES, HOVERS, DARTS ABOVE SURFACE." Box marks: T.P.S.

**Action:** Hovercraft rises, hovers, and darts above the surface as rotating blades inside create air movement to lift the lightweight aluminum body off the ground.

# MAGIC CHOO CHOO

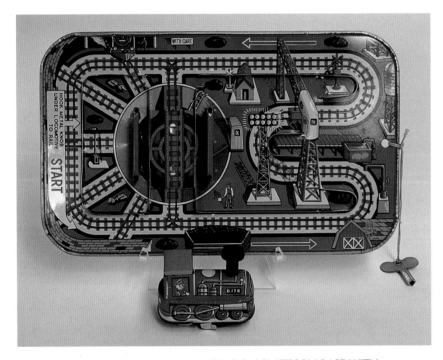

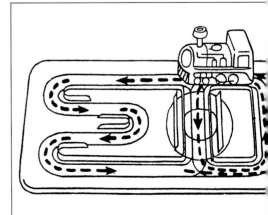

Windup 6cm (2.25in) CHOO CHOO ENGINE ON PLATFORM BASE WITH ROUNDHOUSE AND SWITCHING YARD LITHOGRAPHY, with separate key. Tin with rubber stack on locomotive. Known variations: Base mounted key wind.
Toy Marks: T.P.S. (base, lower right corner).
Size: 9.25in x 5.5in platform (23.5cm x 14cm platform)
Introduced: c.1965. Est. quantity: 40,000 pieces. Scarcity rating: 3.
*Fine to Excellent (C6 to C8): $75-$100, Mint (C10): $125.*

Box text: "MECHANICAL MAGIC CHOO CHOO," "MAGIC ACTION * WINDUP CLOCKWORK * SWITCHING ACTION." Box marks: T.P.S., Franconia.

**Action:** Engine shuttles around platform and changes direction via sliding rail mechanism on roundhouse bridge.

# MAGIC CIRCUS

Windup SKATING 37mm (1.5in) MONKEY AND DANCING 35mm (1.5in) SEAL ON CIRCUS BANDSTAND PLATFORM, with lever key. Tin figures, base, and roof. Plastic platform and pillars.
Toy Marks: T.P.S. (bottom).
Size: H: 6.25in W: 3.25in D3.25in (H: 16cm W: 8cm D: 8cm)
Introduced: c.1968. Est. quantity: 12,000 pieces. Scarcity rating: 3.5.
*Fine to Excellent (C6 to C8): $125-$150, Mint (C10): $175.*

Box text: "2 IN 1 MAGIC CIRCUS," "MAGIC MAGNETIC ACTION * SKATING-DANCING *ART.NO.8121." Box marks: Franconia.

**Action:** Place a figure by center post and pull and release lever. Magnet under platform spins, causing figures to dance and skate around platform.

# MAGIC CROSS ROAD

Windup 7cm (2.75in) LOCOMOTIVE ON PLATFORM BASE WITH COUNTRYSIDE SCENES, BRIDGE, AND WORKING CROSSING GATES, with separate key. Tin with rubber stack on locomotive.
Toy Marks: T.P.S. (upper left side of base), Frankonia (bottom right side of base).
Size: 9.25in x 5.5in platform (23.5cm x 14cm platform)
Introduced: c.1965. Est. quantity: 50,000 pieces. Scarcity rating: 2.
*Fine to Excellent (C6 to C8): $75-$95, Mint (C10): $125.*

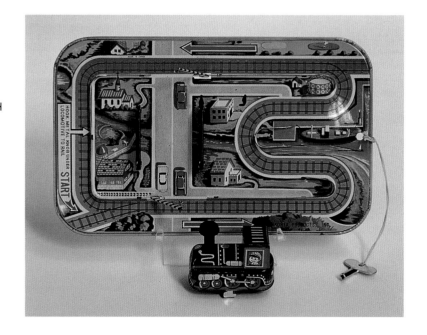

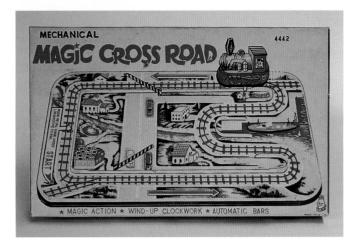

Box text: "MECHANICAL MAGIC CROSS ROAD," "MAGIC ACTION * WIND-UP CLOCKWORK * AUTOMATIC BARS * 4442." Box marks: T.P.S.

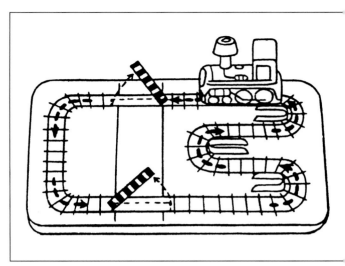

**Action:** Windup steam locomotive goes around platform hooked on guide rail and activates spring loaded railroad crossing gates.

## MAGIC GREYHOUND BUS

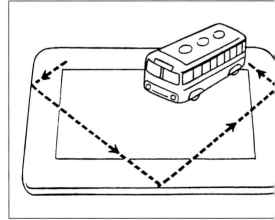

Windup 7.5cm (3in) GREYHOUND BUS ON PLATFORM BASE WITH HIGHWAY SCENES AND TRAVEL POINTS OF CANADA, NEW YORK, FLORIDA, AND SAN FRANCISCO, with fixed key. Tin with rubber wheels on bus.
Toy Marks: T.P.S. (base, lower right corner).
Size: 9.25in x 6.25in platform (23.5cm x 15.5cm platform)
Introduced: c.1967. Est. quantity: 10,000 pieces. Scarcity rating: 4.
Fine to Excellent (C6 to C8): $135-$170, Mint (C10): $210.

Box text: "MECHANICAL MAGIC GREYHOUND BUS * MYSTERY NON-STOP ACTION." Box marks: T.P.S.

**Action:** Bus moves within flat rail around platform base with bump and go action initiated by knob protruding from bottom of bus.

# MAGIC TUNNEL

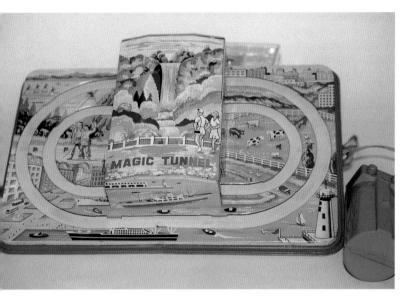

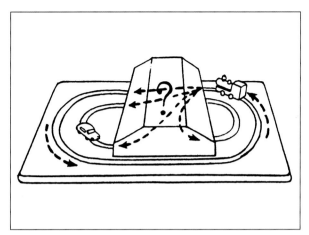

**Action:** Vibrating action causes car and loco to move around track. Car and train have brush type nap underneath to keep them moving forward only. Track changes are made within tunnel making it a mystery surprise as to which track they will exit.

LOCOMOTIVE AND CAR TRAVEL AROUND PLATFORM BASE WITH TRAVEL SCENES AND MAGIC TUNNEL. Battery operated. Tin with plastic train and car. *Photo courtesy of Japan Toys Museum Foundation.*
Size: 9.25in x 6.25in platform (23.5cm x 15.5cm platform)
Introduced: c.1967. Est. quantity: 6,000 pieces. Scarcity rating: 5.
*Fine to Excellent (C6 to C8): $200-$250, Mint (C10): $325.*

# MAMA KANGAROO WITH PLAYFUL BABY IN HER POUCH

Windup JUMPING BABY KANGAROO IN MAMA'S POUCH, with fixed key.
Tin with rubber ears and plastic leaf.
Toy Marks: T.P.S. (rear, left of tail).
Size: H: 6in W: 1.5in D: 2.75in (H: 15cm W: 3.5cm D: 7cm)
Introduced: c.1963. Est. quantity: 20,000 pieces. Scarcity rating: 3.5.
*Fine to Excellent (C6 to C8): $135-$175, Mint (C10): $225.*

Box text: "MECHANICAL MAMA KANGAROO WITH PLAYFUL BABY IN HER POUCH,"
"Wind up with the key, pat baby's head lightly and watch the kangaroo leap." Box marks: T.P.S.

**Action:** Patting the baby's head causes baby to bounce up and down on wire trying to get leaf and berry in mother's mouth.

# MANHATTAN BANK a.k.a. KING KOIN BANK

Windup EMPIRE STATE BUILDING BANK WITH KING KONG LIKE GORILLA CLIMBING EXTERIOR, with fixed key. Plastic. Known variations: also sold as "King Koin Bank" by Leadworks Inc.
Toy Marks: T.P.S. (molded on bottom).
Size: H: 19.75in W: 4in D: 4in (H: 50cm W: 10cm D: 10cm)
Introduced: c.1982. Est. quantity: 30,000 pieces. Scarcity rating: 3.
*Fine to Excellent (C6 to C8): $45-$65, Mint (C10): $80.*

Box text: "MECHANICAL MANHATTAN BANK," "INSERT COIN IN HOLE OF HEAD * GORILLA HOLDS COIN IN HIS MOUTH AND GOES UP WALL OF BUILDING UNTIL THE ROOF-COIN FALLS IN THE BLDG." (In Japanese and English) Box marks: T.P.S.

**Action:** Gorilla carries coin in mouth and climbs the side of the bank. When reaching the top, the coin falls into a slot and gorilla slides back down to repeat action.

# MERCURY EXPLORER

Battery operated BUMP AND GO SPACESHIP WITH ROTATING DOME AND ROTOR BLADES. Tin with plastic dome and rotor blades.
Toy Marks: T.P.S. (left rear of spaceship).
Size: L: 7.75in H: 4.75in W: 4.75in (L: 20cm H: 12cm W: 12cm)
Introduced: c.1970. Est. quantity: 40,000 pieces. Scarcity rating: 3.
*Fine to Excellent (C6 to C8): $175-$225, Mint (C10): $275.*

Box text: "BATTERY OPERATED MAGIC COLOR DOME MERCURY EXPLORER," "NON-STOP ACTION * LIGHT EMANATES OUT THROUGH THE DOME * THE LIGHT CHANGES RED AND BLUE * JET ENGINE SOUND." Box marks: T.P.S.

**Action:** Spaceship moves in bump and go action while dome and rotor blades turn. Light shining through the dome changes back and forth from red to green via reflectors inside.

# MERRY PENGUIN

Windup WADDLING PENGUIN IN PLAID JACKET, with fixed key. Tin.
Toy Marks: T.P.S. (lower left front).
Size: H: 5in W: 2.5in D: 4in (H: 13cm W: 6.5cm D: 10cm)
Introduced: c.1965. Est. quantity: 20,000 pieces. Scarcity rating: 4.5.
*Fine to Excellent (C6 to C8): $250-$300, Mint (C10): $350.*

**Action:** Penguin sways back and forth on its base. This action causes penguin to move about in a waddling manner with jointed arms flapping.

# MERRY PLANE WITH BELL

Windup AMUSEMENT PARK RIDE WITH FLYING PLANES, SPINNERS AND BELL, with fixed key. Plastic with tin bell.
Toy Marks: T.P.S. (molded on top of base).
Size: H: 8in W: 5in D: 5in (H: 20.5cm W: 13cm D: 13cm)
Introduced: c.1981. Est. quantity: 50,000 pieces. Scarcity rating: 2.5.
*Fine to Excellent (C6 to C8): $50-$65, Mint (C10): $75.*

Box text: "MECHANICAL MERRY PLANE WITH BELL" (In Japanese and English). Box marks: T.P.S.

**Action:** 3 different colored planes, attached to tower by rods, fly in circular pattern around tower. Planes increase height by centrifugal force as speed increases. Has ringing bell.

## MERRY-GO-ROUND TRUCK

Battery operated MERRY-GO-ROUND TRUCK WITH MULTICOLOR STYROFOAM BALLS BLOWING IN ENCLOSURE. Plastic. *Photo courtesy of Japan Toys Museum Foundation.*
Toy Marks: T.P.S. (molded in front bumper).
Size: L: 7in H: 6.25in W: 4in (L: 18cm H: 16cm W: 10cm)
Introduced: c.1970. Est. quantity: 10,000 pieces. Scarcity rating: 4.
*Fine to Excellent (C6 to C8): $75-$100, Mint (C10): $125.*

Box text: "BATTERY OPERATED COLOR-FUL BALL BLOWING TOY * MERRY-GO-ROUND TRUCK." Box marks: T.P.S. *Photo courtesy of Japan Toys Museum Foundation*
**Action:** Truck moves in bump and go action with engine sound as balls are blown around inside enclosure. Merry-go-round horses turn as truck moves.

## MICKEY MOUSE CYCLIST

Windup MICKEY MOUSE ON HIGH WHEEL TRICYCLE WITH BELL LITHOGRAPHED WITH HEAD OF MINNIE AND PLUTO, with fixed key. Tin with satin like pants and rubber ears.
Toy Marks: LINEMAR (bell), Walt Disney Productions (bell).
Box text: "MECHANICAL MICKEY MOUSE CYCLIST." Box marks: LINEMAR, Walt Disney Productions.
**Action:** Cycle with ringing bell moves in a wide circle. Mickey's jointed legs move up and down in pedaling action.
Size: H: 7in W: 2.5in L: 4.75in (H: 18cm W: 6.5cm L: 12cm)
Introduced: c.1958. Est. quantity: 20,000 pieces. Scarcity rating: 4.5.
*Fine to Excellent (C6 to C8): $1,500-$2,050, Mint (C10): $2,700.*

Bell detail showing images of Minnie Mouse and Pluto.

# MICKEY MOUSE ROLLER SKATER

Windup ROLLER SKATING MICKEY MOUSE with fixed key. Tin with satin like pants and rubber ears.
Toy Marks: LINEMAR and Walt Disney Productions (paper label, bottom left foot)
Size: H: 6.25in W: 3.25in D: 4in (H: 16cm W: 8cm D: 10cm)
Introduced: c.1958. Est. quantity: 60,000 pieces. Scarcity rating: 3.5.
*Fine to Excellent (C6 to C8): $950-$1,300, Mint (C10): $1,800.*

Box text: "MECHANICAL MICKEY M0USE ROLLER SKATER." Box marks: LINEMAR, Walt Disney Productions. *Dick Rowe Collection.*

**Action:** Mickey skates on wheeled left foot with right leg pushing off, causing body to dip in realistic skating motion. *Dick Rowe Collection.*

# MICKEY MOUSE THE UNICYCLIST

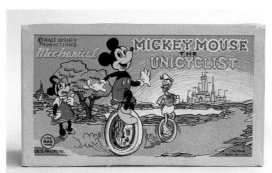

Windup MICKEY MOUSE RIDING ON UNICYCLE, with fixed key. Tin with satin like pants and rubber ears.
Toy Marks: LINEMAR (back, left side), Walt Disney Productions (back, left side).
Size: H: 5.5in W: 2in D: 3.75in (H: 14cm W: 5cm D: 9.5cm)
Introduced: c.1958. Est. quantity: 20,000 pieces. Scarcity rating: 4.
*Fine to Excellent (C6 to C8): $1,000-$1,400, Mint (C10): $1,900.*

Box text: "MECHANICAL MICKEY MOUSE THE UNICYCLIST." Box marks: LINEMAR, Walt Disney Productions.

**Action:** Mickey pedals forward on unicycle then spins around three times and moves forward again in different direction. Lever which extends down from the base lifts one wheel to make toy turn.

## MIDGET LADY-BUG

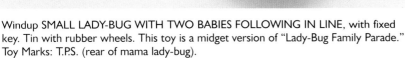

Windup SMALL LADY-BUG WITH TWO BABIES FOLLOWING IN LINE, with fixed key. Tin with rubber wheels. This toy is a midget version of "Lady-Bug Family Parade."
Toy Marks: T.P.S. (rear of mama lady-bug).
Size: L: 7.5in H: 1.25in W: 2.25in (L: 19cm H: 3.5cm W: 5.5cm)
Introduced: c.1967. Est. quantity: 10,000 pieces. Scarcity rating: 4.
*Fine to Excellent (C6 to C8): $90-$115, Mint (C10): $150.*

Box text: "MECHANICAL MIDGET LADY-BUG * No.3792." Box marks: T.P.S., Shackman.

**Action:** Mama lady-bug moves in large circle with two baby lady-bugs following and spinning around.

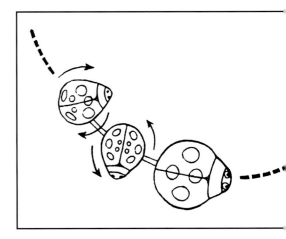

## MIGHTY MIDGET — ANTIQUE MINIATURE AUTO

LEVER ACTION MINIATURE OLD TIME AUTOS. Windup with lever. Tin with plastic tires. These cars also sold in friction version as "Old Fashioned Cars."
Toy Marks: T.P.S. (rear of car) or Shackman (left side of car).
Size: L: 3.25in H: 2.25in W: 2in (L: 8cm H: 6cm W: 5cm)
Introduced: c.1967. Est. quantity: 30,000 pieces. Scarcity rating: 3.
*Fine to Excellent (C6 to C8): $25-$30, Mint (C10): $40.*

Box text: "MIGHTY MIDGET MECHANICAL ANTIQUE MINIATURE CLUTCH LEVER ACTION AUTOS." Box marks: T.P.S. or Shackman.

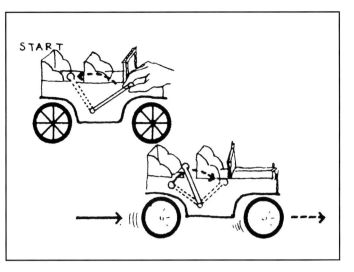

**Action:** Pulling lever back winds mechanism. Releasing it causes car to go forward.

Rear view showing trademarks.

# MIGHTY MIDGET — ASSORTED VEHICLES

MINIATURE VEHICLE ASSORTMENT (like those used on platform toys but without guide rail hooks). Windup with fixed key. Tin with rubber and plastic wheels. Known variations: Tow Truck, Racers, Cement Mixer, Helicopter, Boats, Lumber Truck, Mercedes, Bus, Train, Oil Truck, Tank, Bulldozer, Fire Engine, Taxi, VW, Police, Sports Car.
Toy Marks: Made in Japan (rear).
Size: L: 2.25in (L: 6cm)
Introduced: c.1965. Est. quantity: 200,000 pieces. Scarcity rating: 2.5.
*Fine to Excellent (C6 to C8): $15-$20, Mint (C10): $25.*

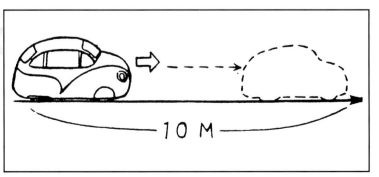

**Action:** Vehicles with powerful motor and gear arrangement capable of traveling 10 meters (40 inches) without re-winding.

Page from 1970 Shackman catalog showing some of the "Mighty Midgets"
Box text: "THE MIGHTY MIDGET * ACTION POWERED * THE WORLD'S TINIEST, MOST POWERFUL MECHANIZED MOTOR-DRIVE." Counter display Box marks: T.P.S., Shackman.

# MIGHTY MIDGET — "ON-TABLE" MECHANICALS

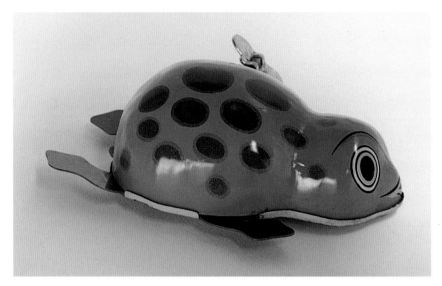

Windup MINIATURE ANIMAL ASSORTMENT THAT WILL NOT FALL OFF TABLE, with fixed key. Tin with plastic legs. Known variations: Frog, Turtle, Bug, Rabbit, Mouse, Squirrel.
Toy Marks: Japan (rear), Shackman (bottom).
Size: L: 3.5in (L: 9cm)
Introduced: c.1965. Est. quantity: 50,000 pieces. Scarcity rating: 2.5.
*Fine to Excellent (C6 to C8): $20-$25, Mint (C10): $30.*

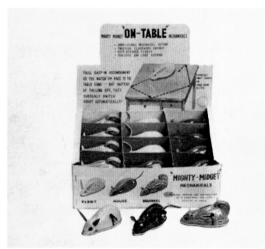

Box text: "MIGHTY MIDGET 'ON-TABLE' MECHANICALS," "YOU'LL GASP IN ASTONISHMENT AS YOU WATCH 'EM RACE TO THE EDGE — BUT INSTEAD OF FALLING OFF THEY SUDDENLY SWITCH ABOUT AUTOMATICALLY." Counter display Box marks: Shackman.

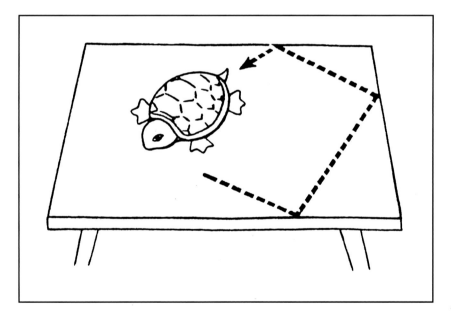

**Action:** Moves forward on one rubber wheel and one non-driven metal wheel. When front of toy starts to drop over edge, rubber bump on bottom stops the toy as driving wheel turns toy in different direction.

# MINI WALKING ROBOT

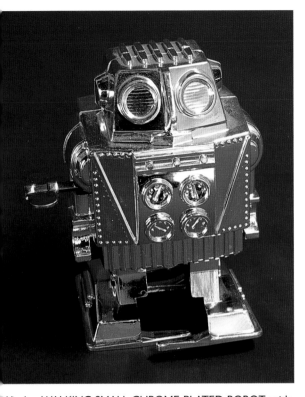

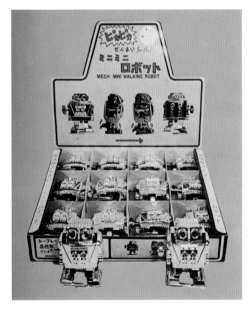

Windup WALKING SMALL CHROME PLATED ROBOT, with fixed key. Chrome plated plastic. Known variations: Black plastic body and chrome robots with white plastic key.
Toy Marks: T.P.S. (molded on rear of head).
Size: H: 3.5in W: 2.25in D: 1.75in (H: 9cm W: 6cm D: 4.5cm)
Introduced: c.1975. Est. quantity: 300,000 pieces. Scarcity rating: 1.5.
*Fine to Excellent (C6 to C8): $15-$20, Mint (C10): $30.*

**Action:** Walks forward one foot at a time with motor noise.

Counter display box text: "MECHANICAL MINI WALKING ROBOT."

# MINIATURE DUNE BUGGY

Windup SMALL DUNE BUGGY THAT DOES WHEELIES, with fixed key. Tin with plastic tires and steering wheel.
Toy Marks: T.P.S. (left rear bumper).
**Action:** Goes forward until fifth wheel on lever arm drops down, causing front wheels to raise in wheelie like action. Arm retracts and dune buggy goes forward again.
Size: L: 3.75in H: 2.25in W: 3in (L: 9.5cm H: 6cm W: 7.5cm)
Introduced: c.1971. Est. quantity: 12,000 pieces. Scarcity rating: 4.
*Fine to Excellent (C6 to C8): $45-$55, Mint (C10): $75.*

# MISSILE ROBOT

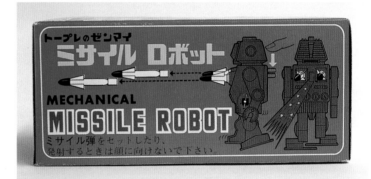

Windup CHROME PLATED WALKING, SPARKING, AND-MISSILE FIRING ROBOT, with fixed key. Plastic with rubber tipped missiles.
Toy Marks: Japan (back of head).
Size: H: 5.5in W: 3.25in D: 2.25in (H: 14cm W: 8.5cm D: 6cm)
Introduced: c.1975. Est. quantity: 60,000 pieces. Scarcity rating: 3.
*Fine to Excellent (C6 to C8): $85-$100, Mint (C10): $125.*

Box text: "MECHANICAL MISSILE ROBOT" (In Japanese and English).
Box marks: T.P.S.

**Action:** Robot walks forward while shooting sparks from behind a small clear chest plate. Spring loaded missiles can be launched from dual switches on back.

# MISSILE TANK

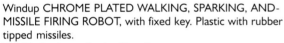

Windup SPRING LOADED MISSILE FIRING TANK, with separate key. Plastic tank, rubber treads, and tin missile platform.
Toy Marks: T.P.S. (rear of missile platform).
Size: L: 6.25in W: 4in H: 7.75in (L: 16cm W: 10cm H: 20cm)
Introduced: c.1971. Est. quantity: 12,000 pieces. Scarcity rating: 4.
*Fine to Excellent (C6 to C8): $85-$110, Mint (C10): $140.*

Box text: "MECHANICAL MISSILE TANK No.5560."
(In Japanese and English). Box marks: T.P.S.

**Action:** Tank moves with on-off lever and makes engine noise. Spring loaded missile platform rotates and adjusts in height. Firing lever launches rubber tipped plastic missile.

# MITSUBISHI ZERO FIGHTER

Windup CHROME PLATED ZERO AIRPLANE WITH SPARKING GUN, with fixed key. Plastic with tin gun and rubber wheels.
Toy Marks: T.P.S. (molded on bottom of left rear wing).
Size: L: 7.5in W: 9in H: 2.5in (L: 19cm W: 23cm H: 6.5in)
Introduced: c.1975. Est. quantity: 50,000 pieces. Scarcity rating: 3.
*Fine to Excellent (C6 to C8): $45-$65, Mint (C10): $75.*

Box text: "MECHANICAL MITSUBISHI ZERO FIGHTER * JAPANESE NAVY TYPE-O CARRIER FIGHTER 'ZEKE'" (In Japanese and English). Box marks: T.P.S.

**Action:** Plane taxis forward with spinning propeller and sparking gun with flint.

# MONKEY BASKET BALL PLAYER

Box text: "MECHANICAL MONKEY BASKET BALL PLAYER WITH BALL & BASKET." Box marks: T.P.S. and Cragstan or T.P.S. only.

**Action:** Monkey bends down and scoops up tin basketball. Arms spring up to make underhand throw into basket. Ball returns and action repeats.

Windup MONKEY SHOOTING BASKETBALL AT BASKET, with fixed key. Tin with netting and tin basketball.
Toy Marks: T.P.S. (back of pants below key).
Size: L: 7.5in H: 7in W: 3.25in (L: 19cm H: 18cm W: 8.5cm)
Introduced: c.1960. Est. quantity: 60,000 pieces. Scarcity rating: 3.
*Fine to Excellent (C6 to C8): $225-$300, Mint (C10): $400.*

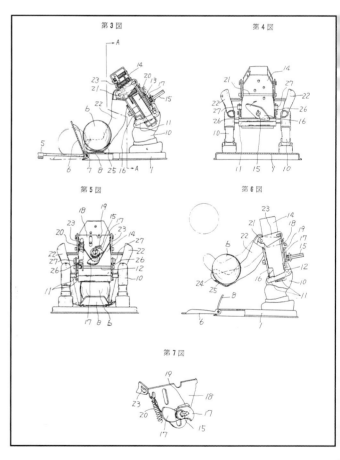

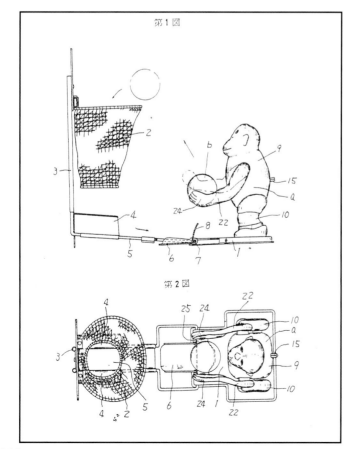

*Courtesy of Toplay (T.P.S.) Ltd.*

# MONKEY GOLFER

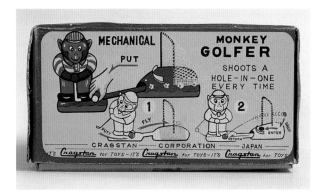

Windup MONKEY HITTING GOLF BALL INTO NET AND HOLE, with fixed key. Tin with netting and rubber ears. Includes two white plastic balls.
Toy Marks: T.P.S. (right rear pant leg), Hirata (left rear pant leg).
Size: H: 4.25in L: 7.75in W: 2.75in (H: 11cm L: 19.5cm W: 7cm)
Introduced: c.1960. Est. quantity: 24,000 pieces. Scarcity rating: 3.5.
Fine to Excellent (C6 to C8): $225-$275, Mint (C10): $325.

Box text: "MECHANICAL MONKEY GOLFER," "SHOOTS A HOLE-IN-ONE EVERY TIME * Cragstan 11666." Box marks: T.P.S., Cragstan.

**Action:** Monkey swings golf club, hitting ball into net which drops into hole and returns. Has Start/Stop switch.

# MOON EXPRESS

Battery operated BUMP AND GO ROCKET SHAPED MOON PASSENGER VEHICLE WITH SPACE PILOT. Tin with plastic front dome, fins, and canopy. Known variations: Unauthorized copies from Taiwan and Hong Kong, not marked T.P.S.
Toy Marks: T.P.S. (rear of left side).
Size: L: 14.25in H: 4.25in W: 3.75in (L: 36cm H: 11cm W: 9.5cm)
Introduced: c.1970. Est. quantity: 60,000 pieces. Scarcity rating: 3.
Fine to Excellent (C6 to C8): $150-$200, Mint (C10): $275.

Box text: "BATTERY OPERATED MAGIC COLOR MOON EXPRESS," "NON-STOP ACTION * INSTANT SPACE NOISE * RED AND GREEN LIGHTS ALTERNATELY APPEAR IN FRONT PART OF THE EXPRESS * YOU CAN OBTAIN BETTER EFFECT OF MAGIC COLOR IN CASE IT IS A SLIGHTLY DARKER ROOM." Box marks: T.P.S.

**Action:** Moon Express moves with bump and go action while front color dome, rear fins, and internal blades rotate. Light shining through the dome changes back and forth from red to green via reflectors inside.

# MOON PATROL

Battery operated BUMP AND GO SPACESHIP WITH ROTOR BLADES AND MOVING COCKPIT WITH ASTRONAUT AND GUN. Tin with plastic dome, rotor blades, and gun.
Toy Marks: T.P.S. (spaceship, behind dome).
Box Text: "BATTERY OPERATED MOON PATROL," "NON-STOP ACTION * ENGINE SOUND * WHIRLING ROTOR BLADE * PILOT IN THE DOME WITH RAY GUNS MOVES RIGHT AND LEFT."
**Action:** Spaceship moves in bump and go action while rotor blades turn. Clear domed cockpit containing astronaut and gun pans left and right in patrolling action.
Size: L: 7.75in H: 4.75in W: 4.75in (L: 20cm H: 12cm W: 12cm)
Introduced: c.1970. Est. quantity: 10,000 pieces.
Scarcity rating: 5.
*Fine to Excellent (C6 to C8): $400-$500, Mint (C10): $625.*

# MOUNTAIN CLIMBER

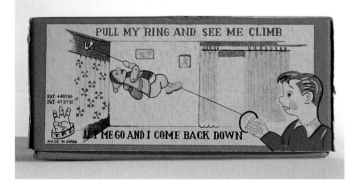

STRING MOUNTAIN CLIMBER WITH BACKPACK. Tin with felt jacket, hat, and backpack and satin like pants.
Toy Marks: T.P.S. (top of right shoe).
Size: H: 7in W: 2.25in D: 2.75in (H: 18cm W: 6cm D: 7cm)
Introduced: c.1962. Est. quantity: 12,000 pieces. Scarcity rating: 3.5.
*Fine to Excellent (C6 to C8): $175-$235, Mint (C10): $300.*

Box text: "MOUNTAIN CLIMBER." Box marks: T.P.S. *Don Hultzman Collection.*

**Action:** Pulling string taut causes figure to climb upward. Relaxing string causes figure to climb downward. String has metal loop at both ends. *Don Hultzman Collection.*

# MOUNTED CAVALRYMAN WITH CANNON

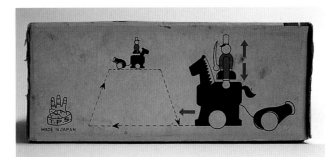

Windup TOY SOLDIER ON HORSE PULLING CANNON, with fixed key. Tin with rubber wheels on horse.
Toy Marks: T.P.S. (left side near rider's foot).
Size: L: 6in H: 5.5in W: 1.75in (L: 15cm H: 14cm W: 4.5cm)
Introduced: c.1960. Est. quantity: 12,000 pieces. Scarcity rating: 3.5.
Fine to Excellent (C6 to C8): $250-$350, Mint (C10): $425.

Box text: "MECHANICAL MOUNTED CAVALRYMAN WITH CANNON," "UP AND DOWN HE GOES ROUND AND ROUND HE GOES AND HIS HORSE WAGS IT'S TAIL." Box marks: SONSCO, T.P.S.

**Action:** Cavalryman moves up and down as horse goes forward with wagging tail. Horse makes right turns via lever which extends from base to raise right wheel from surface.

# MOUSE CHASER

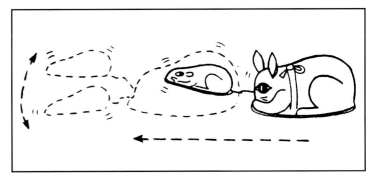

Windup CAT CHASING MOUSE with fixed key. Tin with rubber wheels.
Toy Marks: Made in Japan (rear of cat).
Size: L: 4.25in H: 1.5in W: 1.5in (L: 11cm H: 4cm W: 4cm)
Introduced: c.1967. Est. quantity: 12,000 pieces. Scarcity rating: 3.5.
Fine to Excellent (C6 to C8): $45-$55, Mint (C10): $70.

**Action:** Key wound cat scoots across the surface with small mouse in front. Mouse is attached to wheel on cat by shaped rod, causing it to jump up and forward as they move.

# MOUSE FAMILY

Windup MAMA MOUSE SCURRYING ALONG WITH TWO BABIES FOLLOWING,
with fixed key. Tin with rubber ears on Mama and rubber wheels.
Toy Marks: Japan (rear of Mama mouse), Shackman (base of Mama mouse).
Size: L: 9in H: 1.5in W: 1.5in (L: 23cm H: 3.5cm W: 4cm)
Introduced: c.1967. Est. quantity: 30,000 pieces. Scarcity rating: 3.
*Fine to Excellent (C6 to C8): $35-$45, Mint (C10): $55.*

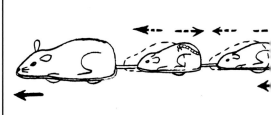

**Action:** Mama mouse with one driven wheel travels in
left circular pattern with babies connected via bar.
Irregular axles on babies cause them to move forward
and backward on bar.

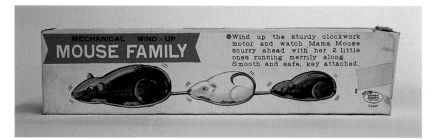

Box text: "MECHANICAL WIND-UP MOUSE FAMILY," "Wind up the
sturdy clockwork motor and watch Mama Mouse scurry ahead with her
2 little ones running merrily along." Box marks: B.Shackman.

# MOUSE RACE CAT

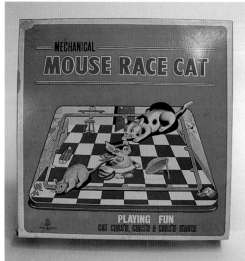

Windup CAT CHASING MOUSE ON PLATFORM BASE WITH
KITCHEN SCENE, with fixed key. Tin.
Toy Marks: T.P.S. (lower left corner of base).
Size: 10.25in x 10.25in platform (26cm x 26cm platform)
Introduced: c.1967. Est. quantity: 30,000 pieces. Scarcity rating: 3.
*Fine to Excellent (C6 to C8): $100-$125, Mint (C10): $150.*

Box text: "MECHANICAL MOUSE RACE CAT," "PLAYING FUN *
CAT CHAS'N, CHAS'N & CHAS'N MOUSE." Box marks: T.P.S.

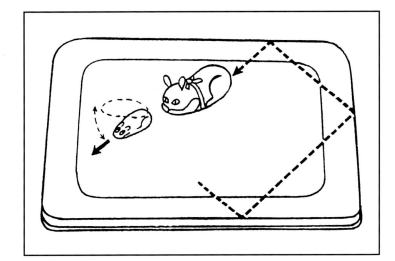

**Action:** Windup cat with attached moving mouse almost getting caught, moves within flat rail around platform base. Bump and go action is initiated by tab protruding from bottom of cat. Key slot in base.

## MR. CATERPILLAR a.k.a. HAPPY CATERPILLAR

Box text: "MECHANICAL MR. CATERPILLAR" OR "MECHANICAL HAPPY CATERPILLAR." Box marks: T.P.S.

**Action:** Caterpillar with spring antennae moves forward as lever movement causes 3-section body to swivel from side to side and spin from friction gear set on wheels.

Windup MOVING AND SWIVELING CATERPILLAR TOY with fixed key. Tin with rubber and plastic wheels.
Toy Marks: T.P.S. (left side behind key).
Size: L: 13in H: 2in W: 2.25in (L: 33cm H: 5cm W: 6cm)
Introduced: c.1960. Est. quantity: 12,000 pieces. Scarcity rating: 3.5.
Fine to Excellent (C6 to C8): $100-$120, Mint (C10): $150.

## MR. SIGNAL

Battery operated WALKING ROBOT LIKE TOY WITH TRAFFIC SIGNAL AS HEAD. Tin with plastic traffic signal, arms, gears, and wheels.
Toy Marks: T.P.S.
**Action:** Mr. Signal moves forward while signal lights go on and off. Gears on chest rotate and hands move up and down.
Size: H: 10.75in W: 4.25in D: 4.25in (H: 27.5cm W: 11cm D: 11cm)
Introduced: c.1970. Est. quantity: 600 pieces. Scarcity rating: 5.
Fine to Excellent (C6 to C8): $700-$900, Mint (C10): $1,150.

# MUSICAL HAND SEWING MACHINE

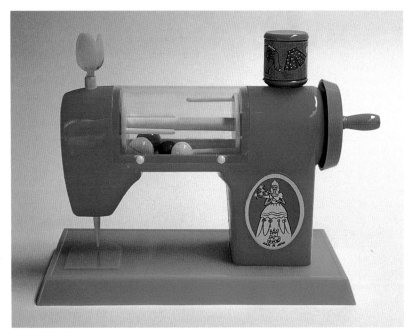

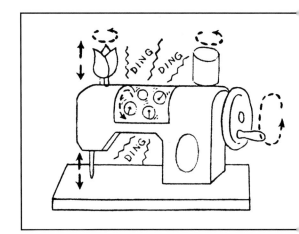

Box text: "MUSICAL HAND SEWING MACHINE PLASTIC ACTION TOY." Box marks: T.P.S.

YOUNG CHILD'S PLASTIC SEWING MACHINE TURNED BY HAND. Plastic.
Toy Marks: T.P.S. (paper label on front and back).
Size: H: 6.5in W: 8.25in D: 4in (H: 16.5cm W: 21cm D: 10cm)
Introduced: c.1967. Est. quantity: 10,000 pieces. Scarcity rating: 5.
*Fine to Excellent (C6 to C8): $60-$75, Mint (C10): $100.*

**Action:** Turning side wheel on sewing machine causes needle to move up and down, spool to rotate, and bells to roll around within visible chamber.

# NEWS SERVICE CAR - WORLD NEWS

Battery operated PORSCHE 911-S WITH WORLD NEWS LITHO AND TV CAMERAMAN ON ROOF. Tin with rubber wheels and plastic cameraman.
Known variations: U-turn Porsche marked TV Service (Red & White), add 20%.
Toy Marks: T.P.S. (dashboard, right side).
Size: L: 10in H: 6in W: 4in (L: 25cm H: 15cm W: 10cm)
Introduced: c.1972. Est. quantity: 20,000 pieces. Scarcity rating: 4.
*Fine to Excellent (C6 to C8): $275-$375, Mint (C10): $475.*

Box text: "BATTERY OPERATED NEWS SERVICE CAR * WORLD NEWS * HIGH-TECHNICAL AUTOMATIC ACTION." Box marks: T.P.S.

**Action:** Car travels forward a short distance then raises up via cam and lever, and makes an approximate 180° turn on two inside wheels. Car lowers and travels forward and action repeats but with a 360° turn.

# OPEN-SHUT BONNET STUNT CAR

Box text: "BATTERY OPERATED OPEN-SHUT BONNET STUNT CAR." Box marks: T.P.S.

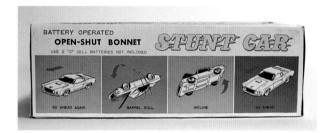

Battery operated MACH 1 1970 FORD MUSTANG BARREL ROLL CAR WITH OPENING HOOD AND FLAME RACE CAR LITHO. Tin with one rubber and three plastic wheels.
Toy Marks: T.P.S. (rear window deck).
Size: L: 10.75in H: 3.25in W: 4.25in (L: 27cm H: 8cm W: 11cm)
Introduced: c.1975. Est. quantity: 60,000 pieces. Scarcity rating: 2.5.
*Fine to Excellent (C6 to C8): $95-$120, Mint (C10): $150.*

**Action:** Car goes forward and raises up on two side wheels via cam and lever mechanism. Car barrel rolls 360° and goes forward again. Front wheels can be positioned for car direction.

# OSCAR THE PERFORMING SEAL (WITH BALL)

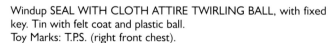

Windup SEAL WITH CLOTH ATTIRE TWIRLING BALL, with fixed key. Tin with felt coat and plastic ball.
Toy Marks: T.P.S. (right front chest).
**Action:** Seal with rotating ball goes forward then stops and raises up on rear flippers, lowers and repeats action. Front flippers move continuously.
Size: H: 6.25in W: 3.75in L: 3.75in (H: 16cm W: 9.5cm L: 9.5cm)
Introduced: c.1959. Est. quantity: 10,000 pieces. Scarcity rating: 3.5.
*Fine to Excellent (C6 to C8): $90-$110, Mint (C10): $140.*

## OSCAR THE SEAL

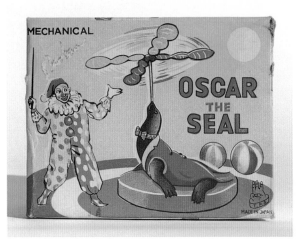

Windup SEAL WITH CLOTH ATTIRE TWIRLING PLASTIC SPINNERS, with fixed key. Tin with felt coat and plastic spinners.
Toy Marks: T.P.S. (right front chest).
Size: H: 7in W: 3.75in L: 3.75in (H: 18cm W: 9.5cm L: 9.5cm)
Introduced: c.1960. Est. quantity: 12,000 pieces. Scarcity rating: 3.5.
*Fine to Excellent (C6 to C8): $75-$100, Mint (C10): $125.*

Box text: "MECHANICAL OSCAR THE SEAL." Box marks: T.P.S.
*Don Hultzman Collection.*

**Action:** Seal with rotating spinners goes forward then stops and raises up on rear flippers, lowers and repeats action. Front flippers move continuously.

## PANGO-PANGO AFRICAN DANCER

Windup LONG NECKED AFRICAN DANCER WITH GRASS SKIRT AND ANKLE CUFFS CARRYING SPEAR & SHIELD, with fixed key. Tin with grass skirt and cuffs with rubber tipped spear.
Toy Marks: T.P.S. (right side of back).
Size: H: 6in W: 2.75in D: 3.5in (H: 15cm W: 7cm D: 9cm)
Introduced: c.1956. Est. quantity: 300,000 pieces. Scarcity rating: 1.5.
*Fine to Excellent (C6 to C8): $125-$175, Mint (C10): $250.*

Box text: "PANGO-PANGO MECHANICAL AFRICAN DANCER." Box marks: T.P.S.

**Action:** Dancer moves neck forward and backward while dancing with vibrating action.

# PAT THE PUP

Window box text: "PAT the PUP WIND-UP." Paper heart shaped tag: "WIND ME UP AND TAP MY NOSE. I will start to move. When I stop, tap my nose. I will bark and start again." Box marks: T.P.S., NGS, Cragstan. Known variations: T.P.S. solid box. *Don Hultzman Collection.*

**Action:** Tapping dog's nose initiates action. Tail spins, body vibrates, and Pat barks via squeak box, as head nods up and down.

Windup SITTING DOG THAT BARKS AND NODS HEAD, with fixed key. Tin body and ears, rubber tail. Toy Marks: T.P.S. (underside, left rear paw), NGS (underside, right rear paw).
Size: H: 5in W: 2.25in D: 4in (H: 13cm W: 5.5cm D: 10cm)
Introduced: c.1965. Est. quantity: 100,000 pieces. Scarcity rating: 2.5.
*Fine to Excellent (C6 to C8): $75-$95, Mint (C10): $125.*

# PERFORMING SEAL AND MONKEY WITH FISH

Windup MONKEY RIDING ON SEAL HOLDS FISH ON POLE, with fixed key. Tin with felt coat and bow on seal. Known variations: Red or blue felt coats.
Toy Marks: T.P.S. (right front chest).
Size: H: 7.75in L: 3.5in W: 3.25in (H: 20cm L: 9cm W: 8cm)
Introduced: c.1960. Est. quantity: 10,000 pieces. Scarcity rating: 4.
*Fine to Excellent (C6 to C8): $300-$400, Mint (C10): $550.*

Coat and tie variation.

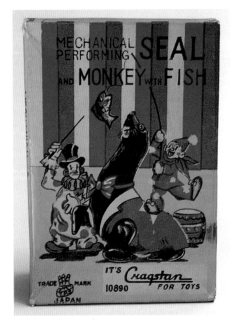

Box text: "MECHANICAL PERFORMING SEAL AND MONKEY WITH FISH Cragstan 10890." Box marks: T.P.S., Cragstan.

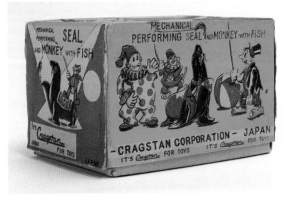

**Action:** Seal rocks up and down on rear flippers trying to reach the fish on the pole.

# PLAYFUL CIRCUS SEALS

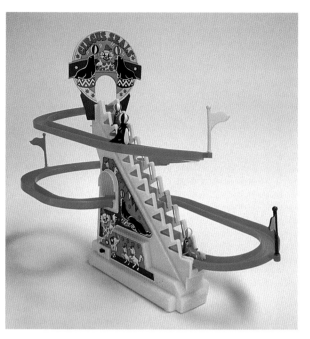

Box text: "BATTERY OPERATED PLAYFUL CIRCUS SEALS™," "The three adorable seals ride while balancing colorful balls on their noses. First they climb up the escalator, then dash down the chute, showing off their acrobatic skills. Action proceeds continuously. All the while the playful seals bark with a 'yip-yip' sound." Box marks: DYToy, Licensed by TOPLAY (T.P.S.) LTD. JAPAN.

Battery operated CIRCUS SEALS CLIMBING STAIRS AND SLIDING DOWN CHUTE. Plastic.
Toy Marks: D.Y.Toy (molded on stairway).
Size: H: 13in L: 16.25in W: 8.25in (H: 33cm L: 41cm W: 21cm)
Introduced: c.1983. Scarcity rating: 2.
*Fine to Excellent (C6 to C8): $25-$35, Mint (C10): $45.*

**Action:** Three seals with balls are lifted up escalator type stairs and sent down winding chute through tunnel to bottom of escalator to repeat action. Complete with sound of barking seals.

# PLAYFUL PUPPY

Bag packaging text: "MECHANICAL PLAYFUL PUPPY * MYSTERY NON-FALL ACTION," marked: T.P.S.

**Action:** Puppy with butterfly attached by wire moves within flat rail around platform base. Bump and go action initiated by knob protruding from bottom of puppy.

Windup 7cm (2.75in) PUPPY CHASING BUTTERFLY ON PLATFORM BASE WITH GARDEN AND ANIMAL SCENES, with fixed key. Tin.
Toy Marks: T.P.S. (bottom right side of base).
Size: 9.25in x 6in platform (23.5cm x 15.5cm platform)
Introduced: c.1964. Est. quantity: 20,000 pieces. Scarcity rating: 4.
*Fine to Excellent (C6 to C8): $125-$150, Mint (C10): $175.*

# PLAYLAND CABLE CAR (Version 1)

Windup 6cm (2.25in) DREAMLAND ANIMAL CABLE CAR SUSPENDED OVER STRAIGHT PLATFORM BASE WITH AMUSEMENT PARK SCENES, with separate key. Tin with rubber wheels on cable car.
Toy Marks: T.P.S. (right side of base between rails).
**Action:** Cable car with wheels moves on ground track to the end where it catches the suspended cable and traverses the cable in the opposite direction via driving wheel located above cable car. When the car reaches the end, it drops to the track and repeats action.
Size: L: 18.25in H: 3.75in W: 1.75in (L: 46.5cm H: 9.5cm W: 4.5cm)
Introduced: c.1960. Est. quantity: 10,000 pieces. Scarcity rating: 5.
*Fine to Excellent (C6 to C8): $550-$750, Mint (C10): $1,000.*

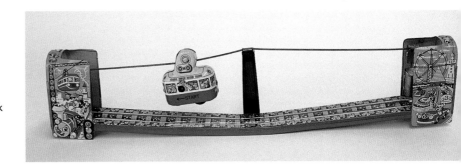

# PLAYLAND CABLE CAR (Version 2)

Box text: "MECHANICAL PLAYLAND CABLE CAR." Box marks: T.P.S.

Windup 6cm (2.25in) DREAMLAND ANIMAL CABLE CAR ON WIRE SUSPENDED OVER CIRCULAR PLATFORM BASE WITH AMUSEMENT PARK SCENES, with separate key. Tin.
Toy Marks: T.P.S. (base, behind one tower).
Size: Diameter: 5.75in H: 4in (Diameter: 14.5cm H: 10cm)
Introduced: c.1965. Est. quantity: 20,000 pieces. Scarcity rating: 4.5.
*Fine to Excellent (C6 to C8): $400-$500, Mint (C10): $650.*

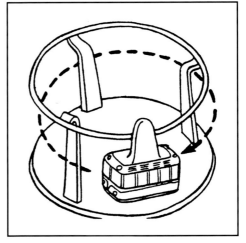

**Action:** Cable car moves around circular wire via driving wheel located above motorized cable car.

Box side panel.

# PLAYLAND SCOOTER

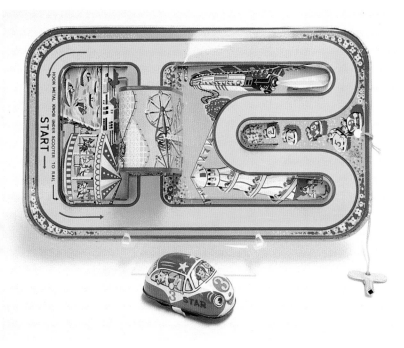

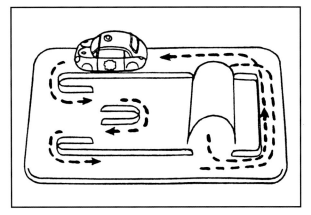

Windup 6cm (2.25in) CAR MARKED "STAR," ON PLATFORM BASE WITH AMUSEMENT PARK SCENES AND TUNNEL, with separate key. Tin.
Toy Marks: T.P.S. (on base, opposite start end).
Size: 9.25in x 5.5in platform (23.5cm x 14cm platform)
Introduced: c.1962. Est. quantity: 30,000 pieces. Scarcity rating: 3.5.
Fine to Excellent (C6 to C8): $125-$145, Mint (C10): $175.

Box text: "MECHANICAL TOY * PLAYLAND SCOOTER ON MAGIC LOOP TRACK." Box marks: T.P.S.

**Action:** Amusement park car runs around platform hooked on guide rail and alternates routes based on sliding rail mechanism in tunnel.

# PLUTO THE UNICYCLIST

Windup PLUTO PEDALING ON UNICYCLE, with fixed key. Tin with rubber tail, ears, and nose.
Toy Marks: LINEMAR and Walt Disney Productions (paper label on bottom of base).
Box text: "MECHANICAL PLUTO THE UNICYCLIST."
Box marks: LINEMAR, Walt Disney Productions.
**Action:** Pluto pedals forward on unicycle which spins around three times and moves forward again in different direction. Lever that extends down from the base lifts one wheel to make toy turn.
Size: H: 5.5in W: 2in D: 2.75in (H: 14cm W: 5cm D: 7cm)
Introduced: c.1958. Est. quantity: 20,000 pieces. Scarcity rating: 4.
Fine to Excellent (C6 to C8): $1,000-$1,400, Mint (C10): $1,800.

# POLICE CAR CHASE

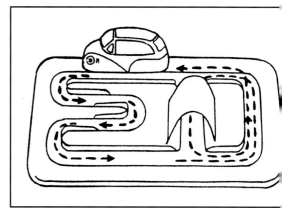

Windup 6cm (2.25in) POLICE CAR ON PLATFORM BASE WITH POLICE SCENES AND TUNNEL, with separate key. Tin.
Toy Marks: T.P.S. (bottom left side of base), Hirata (bottom left side of base).
Size: 9.25in x 5.5in platform (23.5cm x 14cm platform)
Introduced: c.1965. Est. quantity: 30,000 pieces. Scarcity rating: 3.5.
*Fine to Excellent (C6 to C8): $125-$150, Mint (C10): $175.*

Box text: "CRAGSTAN POLICE CAR CHASE * WIND-UP MOTOR * 1561-6." Box marks: Cragstan, T.P.S., NGS.

**Action:** Police car runs around platform hooked on guide rail and alternates routes based on sliding rail mechanism in tunnel.

# POLICE HELICOPTER

Battery operated POLICE HELICOPTER WITH REVOLVING BLADES. Tin with plastic blades and rubber wheels.
Toy Marks: T.P.S. (left side on fuselage).

Box text: "BATTERY OPERATED MYSTERY ACTION POLICE HELICOPTER."

**Action:** Bump and go action with revolving main helicopter blade.
Size: L: 8.25in H: 3.75in W: 2.25in (L: 21cm H: 9.5cm W: 6cm)
Introduced: c.1968. Est. quantity: 30,000 pieces. Scarcity rating: 3.5.
*Fine to Excellent (C6 to C8): $75-$100, Mint (C10): $130.*

Known variations: Police Patrol, Police, Polizei markings.
*Photo courtesy of Tim Hannum.*

# POP EYE PETE

Pop Eye Pete and Comical Clara walking together.

Windup COMICAL WIGGLING BOY WITH POPPING EYES and fixed key. Tin.
Toy Marks: T.P.S. (right rear pants).
Size: H: 5in W: 2.75in D: 2.25in (H: 13cm W: 7cm D: 6cm)
Introduced: c.1967. Est. quantity: 12,000 pieces.
Scarcity rating: 4.
*Fine to Excellent (C6 to C8): $350-$450, Mint (C10): $600.*

Box text: "MECHANICAL POP EYE PETE * THE CUTE KIDS WITH THE COMICAL EYES AND THE WIG WAG WALK." Box marks: T.P.S. *Brynne & Scott Shaw Collection.*

**Action:** Boy wiggles, which causes arms to move like he was walking. Eyes pop in and out in a comical fashion. *Brynne & Scott Shaw Collection.*

# POPEYE AND OLIVE OYL PLAYING BALL

Windup POPEYE AND OLIVE OYL THROWING BALL BACK AND FORTH, with fixed keys. Tin.
Toy Marks: LINEMAR (base of toy), King Features Synd. (base of toy).
Size: L: 19in H: 4.25in W: 1.5in (L: 48cm H: 11cm W: 4cm)
Introduced: c.1958. Est. quantity: 20,000 pieces. Scarcity rating: 4.
*Fine to Excellent (C6 to C8): $1,100-$1,400, Mint (C10): $1,800.*

Box text: "MECHANICAL POPEYE AND OLIVE OYL PLAYING BALL." Box marks: LINEMAR. *Photo courtesy of Herb Smith, Smith House Toys.*
**Action:** Winding each figure causes arms to bounce ball back and forth on wire between Popeye and Olive Oyl. Base folds for box storage.

## POPEYE CYCLIST

Windup POPEYE ON HIGH WHEEL TRICYCLE WITH BELL LITHO-GRAPHED WITH HEAD OF OLIVE OYL AND WIMPY, with fixed key. Tin with satin like pants.
Size: H: 7in W: 2.5in L: 4.75in (H: 18cm W: 6.5cm L: 12cm)
Introduced: c.1957. Est. quantity: 30,000 pieces. Scarcity rating: 3.
*Fine to Excellent (C6 to C8): $1,450-$2,000, Mint (C10): $2,600.*

Box text: "MECHANICAL POPEYE CYCLIST." Box marks: LINEMAR, King Features Synd. *Photo courtesy of Herb Smith, Smith House Toys.*
**Action:** Cycle with ringing bell moves in a wide circle. Popeye's jointed legs move up and down in pedaling action.

Images of Wimpy and Olive Oyl on bell.
Toy Marks: LINEMAR (bell), King Features Synd. (bell).

## POPEYE SKATER

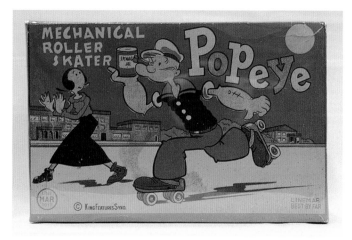

Windup POPEYE ROLLER SKATING WITH PLATE AND CAN OF SPINACH, with fixed key. Tin with satin like pants and rubber wheel.
Toy Marks: LINEMAR (back, left side), King Features Synd. (back).
Size: H: 6.25in W: 3.25in D: 4in (H: 16cm W: 8cm D: 10cm)
Introduced: c.1957. Est. quantity: 60,000 pieces. Scarcity rating: 2.5.
*Fine to Excellent (C6 to C8): $950-$1,350, Mint (C10): $1,800.*

Box text: "MECHANICAL ROLLER SKATER POPEYE J1531." Box marks: LINEMAR, King Features Synd. *Dick Rowe Collection.*

**Action:** Popeye skates on left foot wheel with right leg pushing off, causing body to dip in realistic skating motion. *Dick Rowe Collection.*

# POPEYE THE BASKETBALL PLAYER

Windup POPEYE THROWING BALL UP THROUGH BASKETBALL HOOP, with fixed key. Tin with netting.
Toy Marks: LINEMAR (back of Popeye), King Features Synd. (back of Popeye).
Box text: "MECHANICAL Popeye the Basket Ball Player." Box marks: LINEMAR, King Features Synd.
**Action:** Popeye moves arms up and down, causing ball on wire to bounce up and down through basketball net.
Size: H: 9in W: 2.75in D: 3.5in (H: 23cm W: 7cm D: 9cm)
Introduced: c.1957. Est. quantity: 40,000 pieces. Scarcity rating: 3.
*Fine to Excellent (C6 to C8): $1,500-$2,000, Mint (C10): $2,700.*

# POPEYE UNICYCLIST

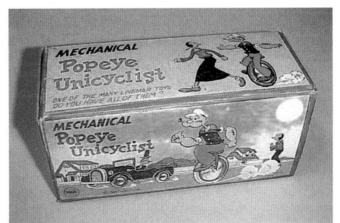

Windup POPEYE PEDALING ON UNICYCLE, with fixed key. Tin with wood pipe and satin like pants.
Toy Marks: LINEMAR (left side of Popeye's back), King Features Synd. (left side of Popeye's back).
Size: H: 6in W: 2in D: 3in (H: 15cm W: 5cm D: 7.5cm)
Introduced: c.1957. Est. quantity: 12,000 pieces. Scarcity rating: 5.
*Fine to Excellent (C6 to C8): $1,600-$2,200, Mint (C10): $3,000.*

Box text: "MECHANICAL POPEYE UNICYCLIST." Box marks: LINEMAR, King Features Syndicate. *Photo courtesy of Randy Ibey, Randy's Toy Shop.*
**Action:** Popeye pedals forward on unicycle which spins around three times and moves forward again in different direction. Lever that extends down from the base lifts one wheel to make toy turn.

## PULL TOY LOCOMOTIVES

TWO TRAIN ENGINES ROTATING ON ROUND BASE WITH
WHEELS, SCENES, BRIDGE, AND TUNNEL. Pull toy. Tin.
Toy Marks: T.P.S. (side of toy).
**Action:** Pulling toy causes trains to move in a circle through the bridge
and tunnel. Wheels on toy are connected to train drive mechanism.
Size: H: 5in Diameter: 6.75in (H: 13cm Diameter: 17cm)
Introduced: c.1965. Est. quantity: 6,000 pieces. Scarcity rating: 4.
*Fine to Excellent (C6 to C8): $150-$190, Mint (C10): $250.*

## PULL-BACK PORSCHE TURBO

Box text: "PULL-BACK PORSCHE TURBO" (In Japanese
and English). Box marks: T.P.S.

Windup MARTINI RACING PORSCHE WITH PULL BACK ACTION. Plastic
with rubber tread on rear wheels. Known variations: Mazda RX-7,
Lamborghini Cauntach, Nissan 280Z.
Toy Marks: T.P.S. (molded on underside).
Size: L: 7.5in W: 3.5in H: 2in (L: 19cm W: 9cm H: 5cm)
Introduced: c.1979. Est. quantity: 30,000 pieces. Scarcity rating: 3.
*Fine to Excellent (C6 to C8): $35-$45, Mint (C10): $55.*

**Action:** As Porsche is pulled back, the wheels wind the
motor allowing it to move forward when released.

## PUSSY CAT CHASING BUTTERFLY a.k.a.
## CASSIUS - THE CAT CHASING BUTTERFLIES

Windup PLUSH CAT CHASING BUTTERFLY ON WIRE, with fixed
key. Plush covered body with tin butterfly.
Toy Marks: None.
Size: L: 8.25in H: 4.75in W: 2.25in (L: 21cm H: 12cm W: 5.5cm)
Introduced: c.1960. Est. quantity: 6,000 pieces. Scarcity rating: 4.5.
*Fine to Excellent (C6 to C8): $150-$190, Mint (C10): $250.*

Box text: "MECHANICAL PUSSY CAT CHASING BUTTERFLY." Box marks: T.P.S. *Don Hultzman Collection.*

Box variation: CASSIUS - THE CAT CHASING BUTTERFLIES, marked Rosko and T.P.S. Add 20%.

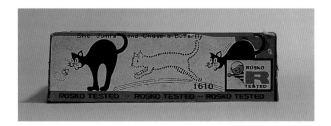

**Action:** Cat with quivering tail and butterfly attached by wire in front of its nose, hunches up twice then leaps forward three times in attempt to catch the butterfly.

# RABBIT AND BEAR PLAYING BALL

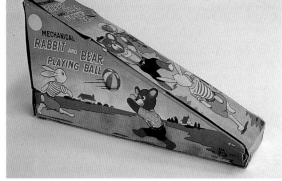

Windup RABBIT AND BEAR THROWING BALL BACK AND FORTH. Toy has two fixed keys, one on each animal. Tin with rubber ears on rabbit and bear.
Toy Marks: T.P.S. (right side of instruction base).
Size: L: 19in H: 4.25in W: 1.5in (L: 48cm H: 11cm W: 4cm)
Introduced: c.1958. Est. quantity: 10,000 pieces. Scarcity rating: 4.
*Fine to Excellent (C6 to C8): $275-$350, Mint (C10): $450.*

Box text: "MECHANICAL RABBIT AND BEAR PLAYING BALL." Box marks: T.P.S.

**Action:** Winding each animal causes arms to bounce ball back and forth on wire between them. Base folds for storage.

# RADAR MISSILE ROBOT

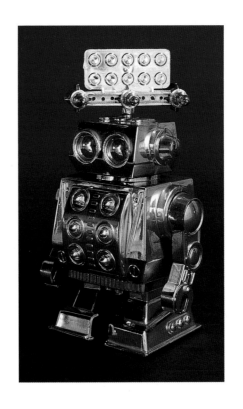

Windup WALKING CHROME PLATED ROBOT WITH ROTATING RADAR SCREEN AND MISSILES ON HEAD, with fixed key. Chrome plated plastic. Known variations: Red/yellow color.
Toy Marks: Made in Japan (lower back of robot body).
**Action:** Robot walks slowly forward as radar antenna with three simulated missiles rotates on top of robot's head. Arms are moveable.
Size: H: 6.75in W: 4in D: 2.75in (H: 17cm W: 10cm D: 7cm)
Introduced: c.1973. Est. quantity: 100,000 pieces. Scarcity rating: 3.5.
*Fine to Excellent (C6 to C8): $50-$60, Mint (C10): $75.*

# RAILROAD SET

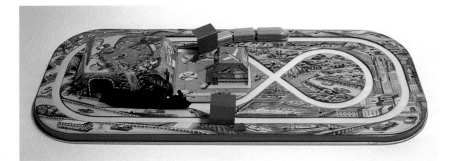

Battery operated WESTERN STEAM TRAIN AND EXPRESS TRAIN TRAVEL AROUND MULTIPLE TRACKS ON PLATFORM BASE WITH CITY AND COUNTRY SCENES, TUNNEL, AND AIRPORT. Tin with plastic trains.
Toy Marks: T.P.S. (lower left side of base), Hirata (lower right side of base).
Size: L: 21.25in W: 9in H: 2.25in (L: 54cm W: 23cm H: 5.5cm)
Introduced: c.1967. Est. quantity: 10,000 pieces. Scarcity rating: 4.
*Fine to Excellent (C6 to C8): $135-$170, Mint (C10): $210.*

Box text: "BATTERY OPERATED RAILROAD SET W/ WESTERN and EXPRESS TRAINS * VIBRATING ACTION." Box marks: T.P.S., Hirata.

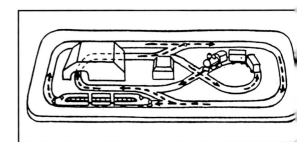

**Action:** Vibrating action causes two three-piece trains to move around track. Trains have brush type nap underneath to keep them moving forward only. Trains change tracks by moving tunnel-like switches. Battery housed in tunnel with on-off switch.

# ROAD RACE

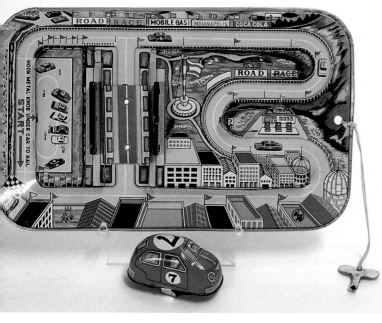

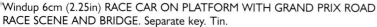

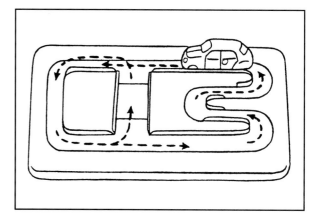

Windup 6cm (2.25in) RACE CAR ON PLATFORM WITH GRAND PRIX ROAD RACE SCENE AND BRIDGE. Separate key. Tin.
Toy Marks: T.P.S. (bottom left side of base), Hirata (bottom right side of base).
Size: 9.25in x 5.5in platform (23.5cm x 14cm platform)
Introduced: c.1965. Est. quantity: 30,000 pieces. Scarcity rating: 3.5.
*Fine to Excellent (C6 to C8): $125-$155, Mint (C10): $200.*

Box text: "MECHANICAL ROAD RACE * MAGIC ACTION * WIND-UP CLOCKWORK * SWITCHING ACTION." Box marks: T.P.S.

**Action:** Race car runs around platform hooked on guide rail and alternates routes based on sliding rail mechanism on bridge.

# ROBOT MACHINE

Battery operated ROBOT MACHINE MOVES THREE 5cm (2in) ROBOTS ON CONVEYOR. Plastic.
Toy Marks: Dah Yang (molded on bottom).
Size: H: 13.5in W: 15.75in D: 7.5in (H: 34cm W: 40cm D: 19cm)
Introduced: c.1984. Scarcity rating: 3.5.
*Fine to Excellent (C6 to C8): $90-$125, Mint (C10): $150.*

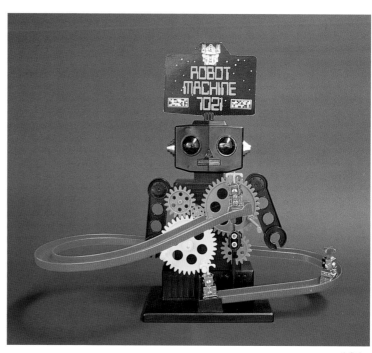

Box text: "Battery Operated KINSMAN ROBOT MACHINE," "Watch as the gears turn, the arms move, the sign rotates and the eyes light up as the robots are moved on their conveyor." Box marks: DY, T.P.S. Licensed by Toplay (T.P.S.) Ltd. Known variations: Japanese version marked T.P.S.

Japanese market version, marked T.P.S. *Japan Toys Museum Foundation Collection.*

**Action:** Gears turn, arms move up and down, sign rotates, and eyes light up. Robots are picked up and transferred by the gear mechanism and placed on the conveyor slide where they return to the bottom to repeat. Disassembles for storage.

## ROLLER SKATING CIRCUS CLOWN

Box text: "BATTERY POWERED REMOTE CONTROL ROLLER SKATING CIRCUS CLOWN." Box marks: T.P.S. *Photo courtesy of John Mendoza*

**Action:** Clown skates on left foot wheel with right leg pushing off, causing body to dip in realistic skating motion. Single C cell battery powers motor in leg.

Battery operated ROLLER SKATING CLOWN WITH WHITE FACE AND WHITE HANDS WITH REMOTE CONTROL BATTERY. Tin with felt jacket and satin finish pants.
Toy Marks: None.
Size: H: 6.25in W: 3.25in D: 4in (H: 16cm W: 8cm D: 10cm)
Introduced: c.1956. Est. quantity: 12,000 pieces. Scarcity rating: 4.5.
*Fine to Excellent (C6 to C8): $800-$1,100, Mint (C10): $1,500.*

# SAMSON THE STRONG MAN

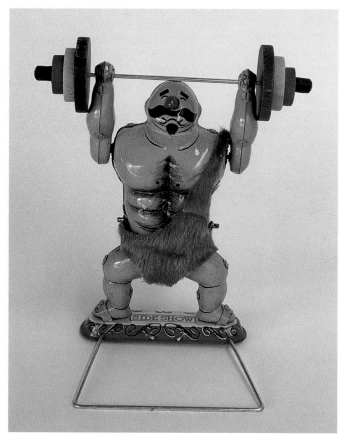

Windup CIRCUS SIDESHOW STRONG MAN LIFTING WEIGHTS ABOVE HEAD, with fixed key. Tin with fur type outfit and wood weights.
Toy Marks: T.P.S. (back right side above key).
Size: H: 6in W: 5in D: 3.25in (H: 15cm W: 13cm D: 8cm)
Introduced: c.1960. Est. quantity: 5,000 pieces. Scarcity rating: 4.5.
*Fine to Excellent (C6 to C8): $550-$700, Mint (C10): $900.*

Box text: "SAMSON THE STRONG MAN." Box marks: T.P.S.

**Action:** Like a real weight lifter, Samson pumps twice then lifts weights above his head then drops them to floor before repeating the actions. Weights are removable.

Box side panel.

# SAND CONVEYOR TRUCK (Version 1)

Friction STEERABLE THREE AXLE SAND CONVEYOR TRUCK WITH LARGE CONVEYOR GEAR MECHANISM.
Tin with rubber wheels and conveyor belt, plastic gear set.
Known variations: See version 2.
Toy Marks: T.P.S. (rear of truck, body and fender), Marusan and Hayashi (truck cab dashboard).
Size: L: 24in H: 7.5in W: 4.25in (L: 61cm H: 19cm W: 11cm)
Introduced: c.1970. Est. quantity: 20,000 pieces. Scarcity rating: 4.5.
*Fine to Excellent (C6 to C8): $250-$300, Mint (C10): $400.*

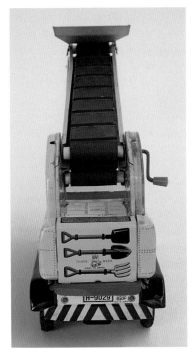

**Action:** Rear wheel friction powered truck. Hand crank turns visible gear set, causing dual conveyor belts to operate. Front wheels can be positioned for direction. Conveyor base slides off truck bed.

Lithography and markings on rear of truck.

## SAND CONVEYOR TRUCK (Version 2)

Friction TWO AXLE SAND CONVEYOR TRUCK WITH SMALL CONVEYOR GEAR MECHANISM. Tin with rubber conveyor and wheels, plastic cab. Known variations: See version Toy Marks: T.P.S. (truck bed in front of conveyor).
Size: L: 15in H: 5in W: 3.75in (L: 38cm H: 13cm W: 9.5cm Introduced: c.1971. Est. quantity: 20,000 pieces. Scarcity rating: 4.5.
Fine to Excellent (C6 to C8): $200-$250, Mint (C10): $325

**Action:** Front wheel friction powered truck with hand crank that turns small visible gear set, causing dual convey belts to operate.

## SATELLITE FLEET

Windup SATELLITE FAMILY OF 3 FLYING SAUCERS AND SPACESHIP, with fixed key. Tin with rubber wheels and plastic radar antenna on spring mount. (Photo missing radar antenna). Known variations: Satellites with star emblems and multi color portholes.
Toy Marks: T.P.S. (right rear side of lead ship).
Size: L: 12.5in H: 2.5in W: 3.25in (L: 32cm H: 6.5cm W: 8cm)
Introduced: c.1964. Est. quantity: 10,000 pieces. Scarcity rating: 4.
Fine to Excellent (C6 to C8): $200-$250, Mint (C10): $325.

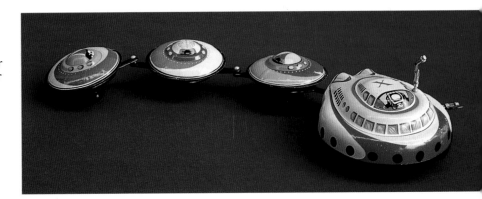

Box text: "MECHANICAL Satellite Fleet." Box marks: T.P.S. *Photo courtesy of Pete Thompson.*

**Action:** Lead space ship moves in alternating directions while 3 saucer type satellites spin as they follow.

## SHUTTLE ZOO TRAIN

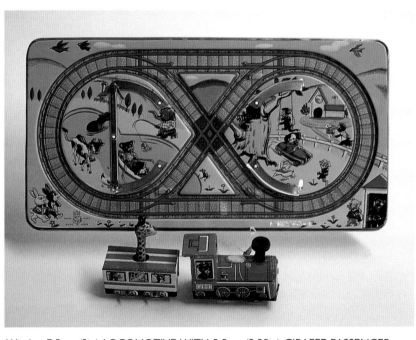

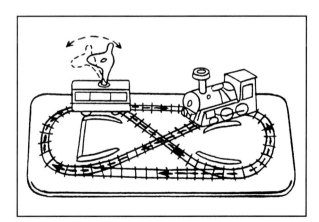

Windup 7.5cm (3in) LOCOMOTIVE WITH 5.5cm (2.25in) GIRAFFE PASSENGER CAR ON PLATFORM BASE WITH ANIMAL COUNTRY SCENES. Fixed key. Tin with rubber smokestack.
Toy Marks: T.P.S. (base, lower left).
Size: 12.5in x 6.25in platform (31.5cm x 16cm platform)
Introduced: c.1962. Est. quantity: 20,000 pieces. Scarcity rating: 3.5.
*Fine to Excellent (C6 to C8): $125-$150, Mint (C10): $200.*

Box text: "SHUTTLE ZOO TRAIN - SLIM GIRAFFE SWINGS HIS LONG NECK." Box marks: T.P.S.

**Action:** Locomotive shuttles around track on platform with automatic rail changing mechanism, leaving giraffe car and picking it up again. As car travels, giraffe swings its neck back and forth.

## SHUTTLING LOCOMOTIVE

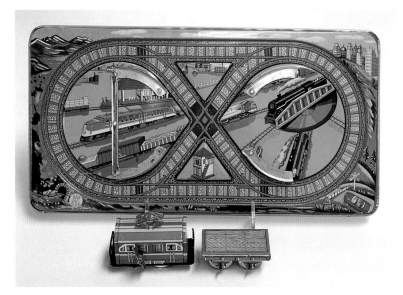

Windup 7cm (2.75in) LOCOMOTIVE WITH PANTOGRAPH AND 5.5cm (2.25in) FLATCAR ON PLATFORM BASE WITH RAILROAD SCENES, with fixed key. Tin.
Toy Marks: T.P.S. (lower left corner of base).
Size: 12.5in x 6.25in platform (31.5cm x 16cm platform)
Introduced: c.1961. Est. quantity: 30,000 pieces. Scarcity rating: 3.5.
*Fine to Excellent (C6 to C8): $125-$150, Mint (C10): $185.*

Box text: "SHUTTLING LOCOMOTIVE * WIND-UP MOTOR." Box marks: T.P.S.

**Action:** Locomotive shuttles around track on platform leaving flatcar and picking it up again with automatic rail change mechanism.

## SHY ANNE INDIAN SKATER

Windup ROLLER SKATING INDIAN WITH BOW & ARROW FROM "BABES IN TOYLAND," with fixed key. Tin with cloth outfit, rubber hands and nose, plastic headdress.
Toy Marks: LINEMAR and Walt Disney Productions (side of right roller skate).
Size: H: 6.75in W: 3.25in D: 4in (H: 17cm W: 8cm D: 10cm)
Introduced: c.1958. Est. quantity: 20,000 pieces. Scarcity rating: 4.
*Fine to Excellent (C6 to C8): $300-$375, Mint (C10): $500.*

Box text: "SHY ANNE MECHANICAL INDIAN SKATER FROM WALT DISNEY'S BABES IN TOYLAND." Box marks: LINEMAR, Walt Disney Productions.

**Action:** Indian skates on wheeled left foot with right leg pushing off, causing body to dip in realistic skating motion.

# SIGHTSEEING BUS

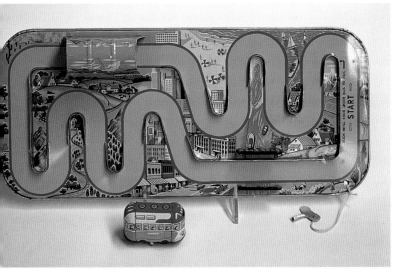

Windup 6cm (2.25in) BUS ON PLATFORM BASE WITH HIGHWAY THROUGH TUNNEL AND CITY, COUNTRY, AND SEASIDE SCENES. Separate key. Tin.
Toy Marks: T.P.S. (left side of platform base).
Size: 15.25in x 6.75in platform (38.5cm x 17cm platform)
Introduced: c.1961. Est. quantity: 60,000 pieces. Scarcity rating: 2.5.
*Fine to Excellent (C6 to C8): $100-$125, Mint (C10): $165.*

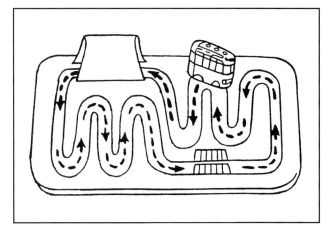

Box text: "SIGHTSEEING BUS * RUNS ON ZIGZAG HIGHWAY * PLACE BUS ON HIGHWAY * HOOK METAL KNOB UNDER BUS TO ROAD GUARD." Box marks: T.P.S.

**Action:** Windup bus goes around platform and through tunnel hooked on zigzag rail.

# SIREN EMERGENCY SERIES

Friction NISSAN JAPANESE AMBULANCE, FIRE CHIEF OR POLICE VAN WITH SIREN SOUND. Tin with plastic wheels.
Toy Marks: T.P.S. (rear or side of vehicle), Ueno (stamped on underside).
Size: L: 6.75in H: 3.5in W: 2.5in (L: 17cm H: 9cm W: 6cm)
Introduced: c.1979. Est. quantity: 60,000 pieces. Scarcity rating: 2.5.
*Fine to Excellent (C6 to C8): $25-$30, Mint (C10): $40.*

Box text: "THE COLLECTION WITH SIREN SOUNDS - AMBULANCE." (In Japanese) Box marks: T.P.S.

Patrol Car box.
**Action:** Friction action with inertia wheel. Motor also connected to small blower. Air movement coupled with pitch changing device creates siren sound.

## SIX TRACK TRAIN SET

STEAM LOCOMOTIVE AND TWO CARS WITH 4 SECTIONS OF CIRCULAR AND 2 SECTIONS OF STRAIGHT INTERLOCKING RAILROAD TRACK. Windup with fixed key. Tin train with plastic frame, wheels, and track.
Toy Marks: Made in Japan (engine and track).
Size: 15in x 18.5in oval track (38cm x 47cm oval track)
Introduced: c.1969. Est. quantity: 12,000 pieces. Scarcity rating: 4.
*Fine to Excellent (C6 to C8): $45-$60, Mint (C10): $75.*

Box text: "MECHANICAL 6 RAILS TRAIN SET" (In Japanese and English). Box marks: T.P.S.

**Action:** Windup 6cm (2.25in) locomotive pulls two cars around a six section interlocking railroad track. Track can be assembled in four piece circle or six piece oval.

## SKATE BOARD

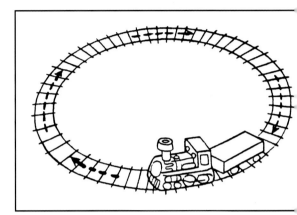

Windup BOY WITH MOVING ARMS, BALANCING ON SKATEBOARD, with fixed key. Plastic. Known variations: Red and blue colors, with and without T.P.S. mark.
Toy Marks: T.P.S. (molded in bottom).
Size: H: 5.25in W: 3.5cm D: 2.25in (H: 13.5cm W: 9cm D: 6cm)
Introduced: c.1979. Est. quantity: 40,000 pieces. Scarcity rating: 3.
*Fine to Excellent (C6 to C8): $35-$45, Mint (C10): $60.*

Box text: "MECHANICAL ROUND AND ROUND SKATE BOARD" (In Japanese and English). Box marks: T.P.S.

**Action:** Boy's arms wave up and down as he balances himself on skate board. Moves forward then turns to right, moves forward then turns to left. Turning action is via levers that drop down to raise one wheel or the other from the surface.

## SKATING CHEF

Windup ROLLER SKATING CHEF WITH FOOD PLATTER, with fixed key. Tin with satin finish pants and apron.
Toy Marks: T.P.S. (back, right side).
Size: H: 6.25in W: 2.5in D: 4in (H: 16cm W: 6.5cm D: 10cm)
Introduced: c.1956. Est. quantity: 100,000 pieces. Scarcity rating: 2.5.
*Fine to Excellent (C6 to C8): $200-$275, Mint (C10): $350.*

Known variation: Dark haired chef, add 20%.

Box text: "MECHANICAL SKATING CHEF * PAT. NO. 53073."
Box marks: T.P.S., HTC.

**Action:** Chef skates on wheeled left foot with right leg pushing off, causing body to dip in realistic skating motion.

## SKATING CHEF - BLACK

Windup ROLLER SKATING BLACK CHEF WITH FOOD PLATTER, with fixed key. Tin with satin finish pants and apron.
Toy Marks: T.P.S. (back, right side).
Size: H: 6.25in W: 2.5in D: 4in (H: 16cm W: 6.5cm D: 10cm)
Introduced: c.1956. Est. quantity: 150,000 pieces. Scarcity rating: 3.
*Fine to Excellent (C6 to C8): $350-$475, Mint (C10): $650.*

**Action:** Chef skates on wheeled left foot with right leg pushing off, causing body to dip in realistic skating motion.

Box text: "MECHANICAL SKATING CHEF * PAT.PEND.NO 5093 & 193813." Box marks: T.P.S., HTC.

## SKIP ROPE ANIMALS

Windup BABY BEAR PLAYS SKIP ROPE WITH SQUIRREL AND DOG, with fixed key. Tin.
Toy Marks: T.P.S. (base, below squirrel).
Size: L: 8in H: 4.75in W: 1.5in (L: 20.5cm H: 12cm W: 4cm)
Introduced: c.1960. Est. quantity: 150,000 pieces. Scarcity rating: 1.5.
*Fine to Excellent (C6 to C8): $125-$155, Mint (C10): $200.*

Box text: "MECHANICAL SKIP ROPE ANIMALS,"
"CRAGSTAN INDUSTRIES 76059." Box marks: T.P.S., Cragstan.

**Action:** Dog's arm rotates rope attached to squirrel, while baby bear jumps over rope. Baby bear is attached to dog's arm by wire.

## SKIPPY, THE TRICKY CYCLIST

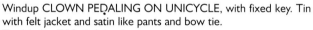

Windup CLOWN PEDALING ON UNICYCLE, with fixed key. Tin with felt jacket and satin like pants and bow tie.
Toy Marks: T.P.S. (left heel).
Size: H: 5.5in W: 2.25in D: 3.25in (H: 14cm W: 6cm D: 8cm)
Introduced: c.1955. Est. quantity: 150,000 pieces. Scarcity rating: 1.5.
*Fine to Excellent (C6 to C8): $160-$200, Mint (C10): $250.*

Box text: "SKIPPY, THE TRICKY CYCLIST MECHANICAL * IT'S FOR TOYS." Box marks: T.P.S., HTC. Known variations: Cragstan box with letters in flags on box.

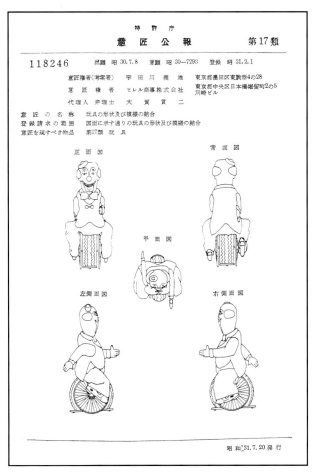

**Action:** Skippy pedals forward on unicycle then spins around three times and moves forward again in different direction. Lever which extends down from the base lifts one wheel to make toy turn.

Box side panel.

Japanese patent applied for in 1955, received in 1956. # 118246. *Courtesy of Toplay (T.P.S.) Ltd.*

# SKULLETON

Windup SKELETON HEAD THAT MOVES AND TURNS, with fixed key. Vinyl with tin base and rubber wheels.
Toy Marks: Made in Japan (underside of jawbone).
Box text: "THE INCREDIBLE WALKING 'SKULLETON' THE SPOOKIEST COOKIEST BONEHEAD RATTLING ABOUT ANYWHERE."
Counter display Box marks: Shackman.
**Action:** Skulleton moves forward on wheels then spins around several times and moves forward again in different direction. Lever which extends down from the base lifts left wheel to make toy turn to the left.
Size: L: 2.75in H: 2.25in W: 1.75in (L: 7cm H: 5.5cm W: 4.5cm)
Introduced: c.1968. Est. quantity: 100,000 pieces. Scarcity rating: 2.5.
*Fine to Excellent (C6 to C8): $40-$50, Mint (C10): $65.*

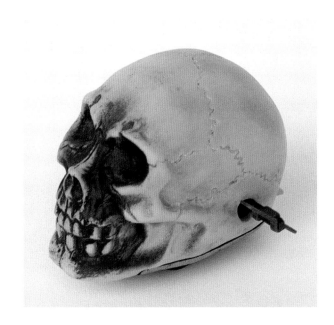

## SKY CIRCUS SPAD XIII

Battery operated SKY CIRCUS SPAD XIII HISPANO-SUIZA BI-PLANE WITH WINGS PERMANENTLY ASSEMBLED. Plastic with tin wings.
Toy Marks: T.P.S. (left tail fin).
Size: L: 9.5in W: 10.5in H: 4.25in (L: 24cm W: 26.5cm H: 11cm)
Introduced: c.1973. Est. quantity: 20,000 pieces. Scarcity rating: 3.
*Fine to Excellent (C6 to C8): $125-$165, Mint (C10): $225.*

Box text: "BATTERY OPERATED SKY CIRCUS SPAD XIII," "AUTO-MATIC ACTION * GO FORWARD * THE TAIL UP AND GO ROUND * GO FORWARD AGAIN." Box marks: T.P.S.

**Action:** Plane with spinning propeller goes forward until lever like third wheel raises the tail of the plane. Different size wheels then cause plane to spin around with nose down and tail up. Plane lowers and repeats actions.

## SLIM THE SEAL & his FRIENDS

Windup CIRCUS PARADE SEAL WITH SPINNERS, PULLING LEAF WAGON WITH 3 SQUIRRELS, with fixed key. Tin with rubber wheels and plastic spinners. Known variations: Juggling Duck is same mold with different lithography.
Toy Marks: T.P.S. (leaf wagon, rear), Japan (bottom of seal).
Size: L: 9.75in H: 6in W: 4.25in (L: 25cm H: 15cm W: 11cm)
Introduced: c.1960. Est. quantity: 12,000 pieces. Scarcity rating: 4.
*Fine to Excellent (C6 to C8): $300-$375, Mint (C10): $500.*

Box text: "Circus Parade 'SLIM' THE SEAL & his FRIENDS. HE JUGGLES AND HUSTLES." Box marks: Cragstan, T.P.S.

**Action:** Seal travels in wide circle while twirling spinners and pulling leaf wagon. Rear two squirrels rotate when wagon moves.

# SNAPPING SHOE

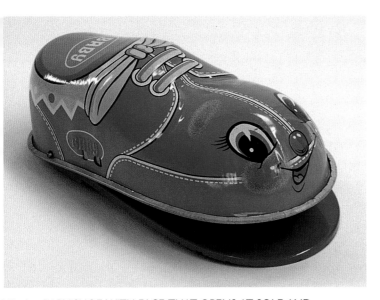

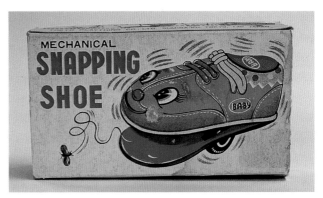

Windup BABY SHOE WITH FACE THAT OPENS AT SOLE AND SNAPS SHUT, with fixed key. Tin with plastic wheels.
Toy Marks: T.P.S. (back left side of shoe).
Size: L: 4.75in H: 1.75in W: 2.25in (L: 12cm H: 4.5cm W: 5.5cm)
Introduced: c.1970. Est. quantity: 10,000 pieces. Scarcity rating: 4.
*Fine to Excellent (C6 to C8): $60-$75, Mint (C10): $100.*

Box text: "MECHANICAL SNAPPING SHOE." Box marks: T.P.S.

**Action:** Shoe moves forward while pins in the wheels cause top of shoe to open and snap shut.

# SOMERSAULT FROG

Box text: "MECHANICAL SOMERSAULT FROG NO.2022." Box marks: T.P.S.

**Action:** Frog moves forward on three wheels then stops as back legs raise the frog off its wheels, causing it to flip in a somersault before repeating action.

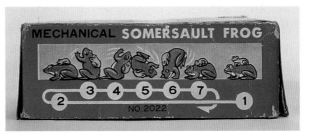

Windup GREEN FROG WITH BABY ON ITS BACK THAT DOES SOMERSAULTS, with fixed key. Plastic with tin base and rubber wheels.
Toy Marks: Made in Japan (molded under right front leg).
Size: L: 4in H: 2.75in W: 3.5in (L: 10.5cm H: 7cm W: 9cm)
Introduced: c.1970. Est. quantity: 15,000 pieces. Scarcity rating: 3.5.
*Fine to Excellent (C6 to C8): $60-$75, Mint (C10): $100.*

## SPACE EXPLORER WITH FLYING SPACE MAN

Friction SPACESHIP WITH LARGE WHEELS AND FLOATING ASTRONAUT. Tin.
Toy Marks: T.P.S. (spaceship body behind left wheel).
Size: L: 6.25in W: 4.75in H: 2.25in + astronaut (L: 16cm W: 12cm H: 6cm+)
Introduced: c.1966. Est. quantity: 10,000 pieces. Scarcity rating: 5.
*Fine to Excellent (C6 to C8): $400-$500, Mint (C10): $650.*

Box text: "FRICTION POWERED SPACE EXPLORER WITH FLYING SPACEMAN." Box marks: T.P.S.

**Action:** Friction movement causes spaceship to move forward and spaceman to circle overhead while connected to spaceship by wire. Spaceship body is the same shape as "Ladybug & Tortoise with Babies."

## SPACE EXPRESS

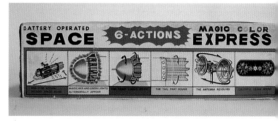

Battery operated BUMP AND GO ROCKET SHAPED, SPACE EXPRESS VEHICLE WITH SPACE PILOT, ROTATING ANTENNA, AND VISIBLE GEARS.
Tin with plastic color dome, fins, antenna, gears, and pilot.
Toy Marks: T.P.S. (rear left side of base).
Size: L: 14.25in H: 4.25in W: 3.75in (L: 36cm H: 11cm W: 9.5cm)
Introduced: c.1971. Est. quantity: 10,000 pieces. Scarcity rating: 5.
*Fine to Excellent (C6 to C8): $550-$700, Mint (C10): $900.*

Box text: "BATTERY OPERATED MAGIC COLOR SPACE EXPRESS," "6-ACTIONS * YOU CAN OBTAIN BETTER EFFECT OF MAGIC COLOR IN CASE IT IS OPERATED IN A SLIGHTLY DARKENED ROOM."
Box marks: T.P.S.

**Action:** Space Express moves with bump and go action while front color dome, rear fins, and external antenna rotate. Light shining through the dome changes back and forth from red to green via reflectors inside. Visible gear sets rotate on either side.

# SPACE STATION CHASE

Battery operated TWO SPACE SHIPS RACING ACROSS UNIVERSE BACKGROUND WITH SLIDE TRACK. Plastic with cardboard backdrop.
Toy Marks: Dah Yang Toy (molded on bottom).
Size: H: 15.75in W: 18.5in D: 10.75in (H: 40cm W: 47cm D: 27cm)
Introduced: c.1997. Scarcity rating: 2.
*Fine to Excellent (C6 to C8): $25-$30, Mint (C10): $35.*

Box text: "SPACE STATION CHASE * Space Ships Race Across the Universe * Mysterious Space Launch Action * Licensed by Toplay (T.P.S.) Ltd.." Box marks: D.Y.Toy, T.P.S.
**Action:** Moving magnet behind space backdrop lifts the space ships and deposits them on downhill slide where they return to repeat actions. Disassembles for storage.

# SPARKING CATERPILLAR BULLDOZER

Windup SPARKING BULLDOZER WITH MOVEABLE BLADE, with fixed key. Tin with rubber treads and plastic wheels, seat, driver, and exhaust stack. Known variations: Japanese version with different litho and exhaust stack.
Toy Marks: T.P.S. (bulldozer blade)
Size: L: 9in H: 5in W: 5in (L: 23cm H: 13cm W: 13cm)
Introduced: c.1971. Est. quantity: 12,000 pieces. Scarcity rating: 4.
*Fine to Excellent (C6 to C8): $100-$125, Mint (C10): $150.*

Box text: "MECHANICAL SPARKING ENGINE CATERPIL-LAR BULLDOZER * STRONG WIND-UP MOTOR." Box marks: T.P.S.
**Action:** Bulldozer moves forward via on-off lever near operator. Internal flint wheel causes sparking which shows through red transparent plastic sides of engine. Blade can be positioned up and down.

## SPARKING ROBOT

Windup WALKING ROBOT SHOOTING SPARKS FROM CHEST, with fixed key. Chrome plated plastic.
Toy Marks: Made in Japan (back of robot body).
**Action:** Robot walks slowly forward as flint rubbing against wheel in body causes sparks to shoot from hole in chest. Arms are moveable.
Size: H: 5.5in W: 2.75in D: 2.75in (H: 14cm W: 7cm D: 7cm)
Introduced: c.1975. Est. quantity: 40,000 pieces. Scarcity rating: 4.
*Fine to Excellent (C6 to C8): $40-$50, Mint (C10): $65.*

## SPINNING PORSCHE

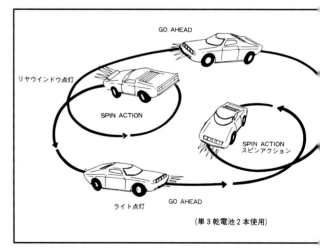

Battery operated MARTINI RACING PORSCHE WITH SPIN OUT ACTION, LIGHT AND GLITTERING CHROME FINISH. Plastic with rubber tread on one wheel. Known variations: Madza RX-7, Lamborghini Cauntach, Lancia Stratos.
Toy Marks: T.P.S. (molded on underside).
Box text: "BATTERY OPERATED SPINNING PORSCHE" (Japanese and English). Box marks: T.P.S.
Size: L: 7.5in W: 3.5in H: 2in (L: 19cm W: 9cm H: 5cm)
Introduced: c.1979. Est. quantity: 40,000 pieces. Scarcity rating: 3.
*Fine to Excellent (C6 to C8): $25-$35, Mint (C10): $45.*

**Action:** As Porsche races in circle, a fifth wheel drops down causing car to move sideways in spin out fashion. Bright interior light shines through rear window.

# SPORTS CAR RACE

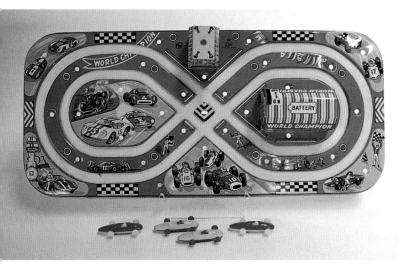

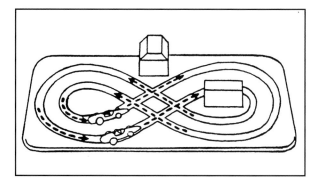

Battery operated 4 SPORTS CARS RACE AROUND FIGURE EIGHT TRACK ON PLATFORM BASE WITH RACE SCENES AND BUILDINGS. Tin platform with plastic cars.
Toy Marks: T.P.S. (upper right center of base), Hirata (upper left center of base).
Size: 14.25in x 5.75in platform (36cm x 14.5cm platform)
Introduced: c.1967. Est. quantity: 15,000 pieces. Scarcity rating: 4.
*Fine to Excellent (C6 to C8): $140-$175, Mint (C10): $225.*

Box text: "BATTERY OPERATED SPORTS CAR RACE * VIBRATING ACTION." Box marks: T.P.S., Hirata.

**Action:** Vibrating action causes cars to move around track. Cars have brush type nap underneath to keep them moving forward only. Battery housed in building with on-off switch.

# SPORTS CYCLE

Battery operated BICYCLE RIDER PEDALING ON SPORT BICYCLE. Plastic with rubber tires and vinyl rider.
Toy Marks: None.
Size: L: 7.75in H: 6in W: 3in (L: 19.5cm H: 15cm W: 7.5cm)
Introduced: c.1975. Est. quantity: 10,000 pieces. Scarcity rating: 4.
*Fine to Excellent (C6 to C8): $150-$190, Mint (C10): $250.*

Box text: "BATTERY OPERATED Go' Go' Young SPORTS CYCLE." Box marks: T.P.S.

**Action:** Bicycle moves in direction of adjustable front wheel. Motor drives rear tire via friction axle. Rear tire is connected to pedals with rubber belt, causing rider to move as if pedaling. Rider is jointed at knee and hip.

## STANDING TURTLE

Windup TURTLE STANDING ON BACK LEGS with fixed key. Tin with vinyl head. *Japan Toys Museum Foundation Collection.*
Toy Marks: T.P.S. (bottom rear of shell).
**Action:** Turtle sways back and forth causing its stationary base to move in walking manner while its front legs move up and down. Makes squeaking noise.
Size: H: 6.25in W: 4.75in D: 3.25in (H: 16cm W: 12cm D: 8cm)
Introduced: c.1968. Est. quantity: 1,200 pieces. Scarcity rating: 5.
*Fine to Excellent (C6 to C8): $150-$200, Mint (C10): $275.*

## STOP AND GO TRUCK SERIES

Box text: "MINIATURE MECHANICALS * STOP AND GO TRUCK SERIES." Box marks: T.P.S., F.E.White.
**Action:** Truck goes in circular fashion via angled front wheels. Lever raises back wheels causing truck to stop. Apparatus on truck is then raised and lowered by second lever before repeating actions.

Windup STOP AND GO FIRE, DUMP, AND CRANE TRUCKS THAT RAISE AND LOWER THEIR APPARATUS, with fixed key. Plastic with tin base. Known variations: Earliest version had ringing bell and rubber tires.
Toy Marks: None.
Size: L: 4.25in H: 3.25in W: 2.25in (L: 11cm H: 8cm W: 6cm)
Introduced: c.1968. Est. quantity: 30,000 pieces. Scarcity rating: 3.
*Fine to Excellent (C6 to C8): $25-$35, Mint (C10): $50.*

# STUNT PLANE

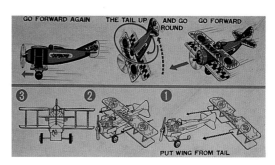

Battery operated STUNT SPAD XIII BI-PLANE S-7, WITH REMOVABLE WINGS.
Plastic with tin wings.
Toy Marks: T.P.S. (left tail fin).
Size: L: 9.5in W: 10.5in H: 4.25in (L: 24cm W: 26.5cm H: 11cm)
Introduced: c.1974. Est. quantity: 20,000 pieces. Scarcity rating: 3.
*Fine to Excellent (C6 to C8): $125-$160, Mint (C10): $215.*

Box text: "BATTERY OPERATED STUNT PLANE *
AUTOMATIC ACTION." Box marks: T.P.S.

**Action:** Plane with spinning propeller goes forward
until lever like third wheel raises the tail of the plane.
Different size wheels then cause plane to spin around
with nose down and tail up. Plane lowers and repeats
actions. Wings remove for storage.

# SUPER FLYING HELICOPTER

Battery operated HIGHWAY PATROL HELICOPTER WITH TIN
WINDSHIELD AND SIMULATED TAKE-OFF AND LANDING
ACTION. Tin with plastic propellers.
Toy Marks: T.P.S. (tail, right side), T.P.S. (molded on end of main blade).
Size: L: 13.75in W: 4.25in H: 5in (L: 35cm W: 11.5cm H: 13cm)
Introduced: c.1969. Est. quantity: 24,000 pieces. Scarcity rating: 3.5.
*Fine to Excellent (C6 to C8): $175-$225, Mint (C10): $275.*

Box text: "SUPER FLYING HELICOPTER," "BATTERY
POWERED * AUTOMATIC ACTIONS * REVOLVING
PROPELLER * TAKE-OFF AND LANDING * GO-STOP *
FLYING AROUND * BLINKING LIGHTS * REAL ENGINE
NOISE * DYNAMIC ACTION." Box marks: T.P.S.

**Action:** Helicopter with revolving blades, blinking lights,
and engine noise, moves forward. Pedestal lowers from
helicopter to raise it up from ground in lift off fashion.
Helicopter spins as if flying, then lowers to repeat action.

３

の作動を反覆すれば亦機体は前述の作動を行うのである、即ち走行を停止すれば機体は上昇し、定位置に達すると機体を前下りに傾斜すると同時に廻動し、これが停止すると降下し走行を繰返すのである。

本考案は以上のような簡単な装置によつて従来のヘリコプター玩具に見ない作動を行うものにしてこの作動は真実のヘリコプターの作動に似ており、從てヘリコプターの実感がよくあらわれて子供等に大きな感興を与える効果がある。

本考案は単にヘリコプターに限らず宇宙船玩具その他垂直昇降を行なわせる玩具類に利用することもできるのである。

**実用新案登録請求の範囲**

下端に支持台板を取付けた支柱を機体の底板の

４

適所に穿設した透孔に嵌合させ、底板上に原動機及び伝動機構を装備した機枠を取付け、機枠の下端に横に貫設される駆動軸の両端に取付けた車輪を底板下に適宜臨ませ、機枠の外側適所に後端を軸支した作動枠を装置し、これが前端に支枠を軸支し、この支枠に支柱の上端を軸着し、支柱の支軸を弧状に形成した機枠の前端縁に沿つて移動するようになし、弧状前端縁の下端をU字形溝に形成し、該U字形溝に近い適所に中間軸上に取付けたピニオンを配設し、これに係脱する歯車を支柱の上端部に取付け、作動枠の適所を機枠の１側に突出した廻転軸の一端に取付けるクランクと連結し、機枠に設けた駆動縦軸上端にプロペラーを取付けたヘリコプター玩具の作動装置。

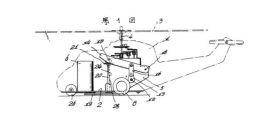

第 1 図

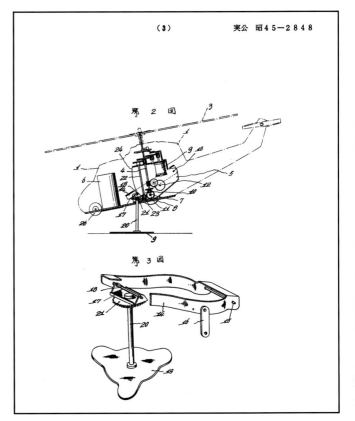

第 2 図

第 3 図

# SUPER FLYING POLICE HELICOPTER

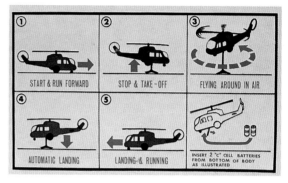

Battery operated HIGHWAY PATROL POLICE HELICOPTER WITH PLASTIC WINDSHIELD AND SIMULATED TAKE-OFF AND LANDING. Tin with plastic propellers and cockpit windshield.
Toy Marks: T.P.S. (tail, right side), T.P.S. (molded on end of main blade).
Size: L: 13.75in W: 4.25in H: 5in (L: 35cm W: 11.5cm H: 13cm)
Introduced: c.1971. Est. quantity: 40,000 pieces. Scarcity rating: 3.
*Fine to Excellent (C6 to C8): $150-$200, Mint (C10): $250.*

Box text: "SUPER FLYING POLICE HELICOPTER," "BATTERY POWERED * AUTOMATIC ACTIONS * REVOLVING PROPELLER * TAKE-OFF AND LANDING * GO-STOP * FLYING AROUND * BLINKING LIGHTS * REAL ENGINE NOISE * DYNAMIC ACTION * NO.9094." Box marks: T.P.S.

**Action:** Helicopter with revolving blades, blinking lights, and engine noise moves forward. Pedestal lowers from helicopter to raise it up from ground in lift off fashion. Helicopter spins as if flying then lowers to repeat action.

# SUPER FLYING TRAFFIC CONTROL HELICOPTER

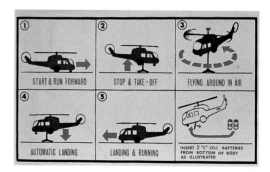

Battery operated TRAFFIC CONTROL HELICOPTER WITH PLASTIC WIND-SHIELD AND SIMULATED TAKE-OFF AND LANDING. Tin with plastic propellers and cockpit windshield.
Toy Marks: T.P.S. (tail, right side), T.P.S. (molded on end of main blade).
Size: L: 13.75in W: 4.25in H: 5in (L: 35cm W: 11.5cm H: 13cm)
Introduced: c.1970. Est. quantity: 24,000 pieces. Scarcity rating: 3.5.
*Fine to Excellent (C6 to C8): $175-$225, Mint (C10): $275.*

Box text: "SUPER FLYING TRAFFIC CONTROL HELICOPTER," "BATTERY POWERED * AUTOMATIC ACTIONS * REVOLVING PROPELLER * TAKE-OFF AND LANDING * GO-STOP * FLYING AROUND * BLINKING LIGHTS * REAL ENGINE NOISE * DYNAMIC ACTION * NO.9094." Box marks: T.P.S.

**Action:** Helicopter with revolving blades, blinking lights, and engine noise moves forward. Pedestal lowers from helicopter to raise it up from ground in lift off fashion. Helicopter spins as if flying then lowers to repeat action.

## SUPER THUNDER WOLF

Battery operated BELL 206B THUNDER WOLF HELICOPTER THAT CONVERTS TO FIGHTER. Plastic.
Toy Marks: DY (molded on battery cover).
Size: L: 19.25in H: 6in W: 4in (L: 49cm H: 15cm W: 10cm)
Introduced: c.1985. Scarcity rating: 2.5.
*Fine to Excellent (C6 to C8): $25-$35, Mint (C10): $45.*

Box text: "SUPER THUNDER WOLF," "TWO TOYS IN ONE * BELL 206B HELICOPTER * 1/24 SCALE ACTION MODEL * 6 FUNCTIONS * LICENSED BY TOPLAY (T.P.S.)LTD." Box marks: DAH YANG INDUSTRIAL TOY CO. LTD.
**Action:** Helicopter with revolving blades, blinking lights, and engine noise moves forward. Pedestal lowers from helicopter to raise it up from ground in lift off fashion. Helicopter spins as if flying then lowers to repeat action. Front wheels positionable. Copter converts by replacing running skids with laser cannons.

## SUSIE THE OSTRICH

Windup OSTRICH PULLING CLOWN IN CART, with fixed key. Tin with rubber legs and cloth ribbon reins.
Toy Marks: T.P.S. (rear of clown driver).
Size: L: 6.25in H: 5in W: 2.75in (L: 16cm H: 13cm W: 7cm)
Introduced: c.1960. Est. quantity: 10,000 pieces. Scarcity rating: 4.5.
*Fine to Excellent (C6 to C8): $600-$750, Mint (C10): $950.*

Box text: "MECHANICAL SUSIE THE OS-TRICH." Box marks: T.P.S.

**Action:** Neck moves back and forth as ostrich walks. Spring loaded clown driver with whip rod rocks on seat while holding reins attached to ostrich neck.

# SUZY BOUNCING BALL a.k.a.
## PRETTY LITTLE GIRL WITH PONYTAIL PLAYING BOUNCING THE BALL

Windup GIRL WITH PONYTAIL BOUNCING BALL WITH HAND, with fixed key. Tin with vinyl head.
Toy Marks: T.P.S. (top of ball near wire).
Size: H: 5.5in W: 2.75in D: 4.75in (H: 14cm W: 7cm D: 12cm)
Introduced: c.1960. Est. quantity: 50,000 pieces. Scarcity rating: 2.
*Fine to Excellent (C6 to C8): $95-$125, Mint (C10): $150.*

Box text: "MECHANICAL SUZY BOUNCING BALL." Box marks: T.P.S. Box variation: "PRETTY LITTLE GIRL WITH PONYTAIL PLAYING BOUNCING THE BALL - CRAGSTAN -ANIMATED WINDUP," add 15%.

**Action:** Girl bounces ball between floor and her right hand. Both arms move as ball moves up and down attached to body mechanism by wire.

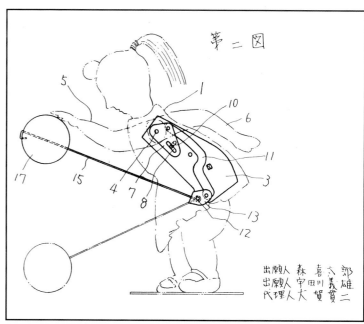

Outline and mechanical detail drawing. *Courtesy of Toplay (T.P.S.) Ltd.*

# TAKE OFF AIRPORT

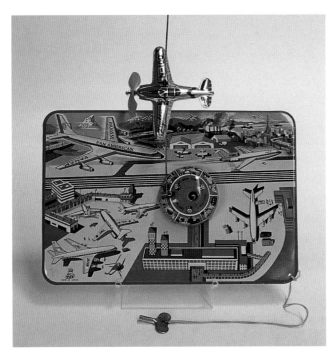

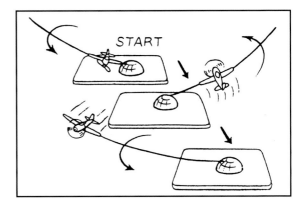

Windup 8cm (3.25in) PLANE MARKED FIGHTER N-100 FLIES OVER PLATFORM BASE WITH AIRPORT SCENE AND CONTROL TOWER. Separate key. Tin with plastic propeller on plane.
Toy Marks: T.P.S. (bottom left side of base).
Size: 9.25in x 5.5in platform (23.5cm x 14cm platform)
Introduced: c.1965. Est. quantity: 30,000 pieces. Scarcity rating: 3.5.
*Fine to Excellent (C6 to C8): $125-$155, Mint (C10): $200.*

Box text: "WIND-UP TAKE-OFF AIRPORT * PLANE TAKES OFF AND LANDS * 6166." Box marks: T.P.S., Franconia.

**Action:** Plane attached to control tower by shaped guide wire takes off and lands via centrifugal action, while circling airport.

# TANK ON A BATTLEFIELD

Box text: "TANK ON A BATTLEFIELD * WIND UP MOTOR." Box marks: T.P.S., Cragstan. Box also seen with T.P.S. markings only.

Windup 6cm (2.25in) MILITARY TANK ON PLATFORM BASE WITH BATTLEFRONT SCENES, with separate key. Tin.
Toy Marks: T.P.S. (bottom right side of base).
Size: 9.25in x 5.5in platform (23.5cm x 14cm platform)
Introduced: c.1964. Est. quantity: 30,000 pieces. Scarcity rating: 3.
*Fine to Excellent (C6 to C8): $100-$125, Mint (C10): $160.*

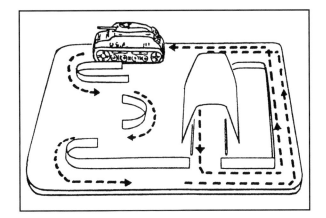

**Action:** Battle tank runs around platform hooked on guide rail and alternates routes based on sliding rail mechanism in tunnel.

# TETSUJIN 28 GO (GIGANTOR)

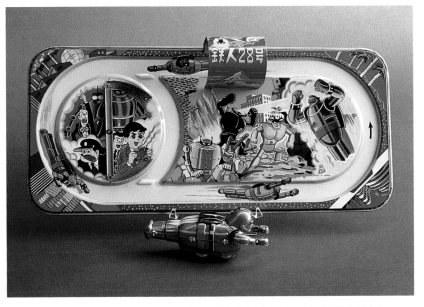

Box text: "MECHANICAL RUNNING TETSUJIN 28TH" (In Japanese). Box marks: T.P.S.

**Action:** Tetsujin 28 runs around track and through tunnel on platform in small or large circular pattern determined by automatic rail changing mechanism.

Windup 12.5cm (5in) CARTOON TETSUJIN 28 IRONMAN ROBOT ON PLAT-FORM BASE WITH ASSOCIATED CARTOON CHARACTERS. Separate key. Tin. [Tetsujin 28: Iron body robot with the head of a knight, controlled by a boy, battles evil. By Mitsuteru Yokoyama.]
Toy Marks: T.P.S. (right center of base).
Size: L: 15.25in W: 6.75in H: 2.25in (L: 38.5cm W: 17cm H: 6cm)
Introduced: c.1965. Est. quantity: 50,000 pieces. Scarcity rating: 4.
*Fine to Excellent (C6 to C8): $1,250-$1,700, Mint (C10): $2,200.*

# TETSUWAN MIGHTY ATOM (ASTROBOY)

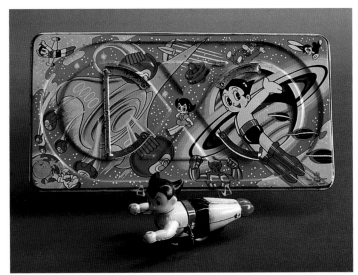

Windup 12cm (4.75in) ASTROBOY ON PLATFORM BASE WITH ASTROBOY CHARACTER AND SPACE SCENES, with fixed key. Tin with vinyl head. [Tetsuwan (Iron Arm) Atom is the creation of Osama Tezuka.]
Toy Marks: T.P.S. (lower right corner of base).
Size: 12.5in x 6.25in platform (31.5cm x 16cm platform)
Introduced: c.1965. Est. quantity: 30,000 pieces. Scarcity rating: 5.
*Fine to Excellent (C6 to C8): $1,300-$1,750, Mint (C10): $2,300.*

Box text: "RUNNING TETSUWAN MIGHTY ATOM * AUTOMATIC RAIL CHANGE" (In Japanese). Box marks: T.P.S.

**Action:** Astroboy runs around track on platform in circular or figure eight pattern determined by automatic rail changing mechanism.

# TETSUWAN ATOM (ASTROBOY) FLYING SAUCER FAMILY

Friction TETSUWAN ATOM (ASTROBOY) ON LEAD SAUCER FOL-LOWED BY CHARACTERS OF PROFESSOR OCHANOMIZU, URAN AND MR. HIGEOYAJI. Tin with vinyl head and rubber wheels.
Toy Marks: Tada - Atom (left rear side of lead saucer).
Size: L: 12.25in H: 3.25in W: 3.25in (L: 31cm H: 8.5cm W: 8cm)
Introduced: c.1965. Est. quantity: 30,000 pieces. Scarcity rating: 5.
*Fine to Excellent (C6 to C8): $1,000-$1,400, Mint (C10): $1,900.*

Astroboy leading his family.

Box text: "TETSUWAN ATOM FLYING SAUCER FAMILY" (In Japanese). Box marks: Tada-Atom. Tetsuwan (Iron Arm) Atom is the creation of Osama Tezuka.

**Action:** Atom saucer moves forward via friction movement with three connected and spinning saucers following. Saucers spin via friction contact with wheels.

# THE HUSTLER

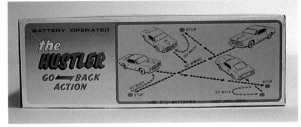

Battery operated RED, WHITE AND BLUE 1970 MUSTANG "THE HUSTLER" WITH RACING LOGOS AND BLUE TINTED WINDOWS. Tin with plastic tires and chassis.
Toy Marks: T.P.S. (top of left rear fender).
Size: L: 10.75in H: 3.25in W: 4.25in (L: 27cm H: 8cm W: 11cm)
Introduced: c.1977. Est. quantity: 40,000 pieces. Scarcity rating: 2.5.
*Fine to Excellent (C6 to C8): $100-$125, Mint (C10): $150.*

Box text: "BATTERY OPERATED the HUSTLER * GO BACK ACTION." Box marks: T.P.S.

**Action:** Goes ahead, stops, reverses, and turns with squealing tire sound. Goes ahead again in an action similar to a demolition derby car.

# THE HUSTLER MUSTANG HANDLING TEST CAR

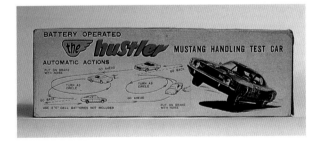

Battery operated BLUE TONE 1970 MACH 1 RACING MUSTANG MARKED "7 - the HUSTLER," WITH ADJUSTABLE FRONT WHEELS, OPEN WINDOWS, AND VISIBLE INTERIOR. Tin with plastic tires and steering wheel.
Toy Marks: T.P.S. (rear window deck).
Size: L: 10.75in H: 3.25in W: 4.25in (L: 27cm H: 8cm W: 11cm)
Introduced: c.1971. Est. quantity: 30,000 pieces. Scarcity rating: 3.5.
*Fine to Excellent (C6 to C8): $140-$175, Mint (C10): $225.*

Box text: "BATTERY OPERATED the HUSTLER MUSTANG HANDLING TEST CAR."
Box marks: T.P.S.

**Action:** Car goes ahead and stops with squealing brake sound. Car spins out in tight circle via underside pivoting gear and wheel mechanism then backs up before going ahead again and repeating actions. Front wheels can be positioned for direction.

Underside detail of pivoting gear and wheel mechanism.

# THE MONKEY TREE

Battery operated MONKEYS CLIMBING TREE AND COMING DOWN HAND OVER HAND.
Toy Marks: Dah Yang (molded on bottom).
Size: H: 19.25in W: 11in D: 8.75in (H: 49cm W: 28cm D: 22.5cm)
Introduced: c.1997. Scarcity rating: 1.5.
*Fine to Excellent (C6 to C8): $20-$25, Mint (C10): $30.*

Box text: "THE MONKEY TREE," "LET MONKEY AROUND * Monkeys swing with a silly hand over hand Monkey Motion * Licensed by Toplay (T.P.S.) Ltd." Box marks: D.Y.TOY.

**Action:** Monkeys are lifted up tree trunk as if climbing and moved on to rail where they start down hand over hand. As they near the bottom they drop to slide which runs to base of the tree before starting over again.

# THE SWINGER (Version 1)

Battery operated LIME GREEN RACING MACH 1 1970 MUSTANG. MARKED "THE SWINGER," WITH ADJUSTABLE FRONT WHEELS, OPEN WINDOWS, AND VISIBLE INTERIOR. Tin with plastic tires and steering wheel.
Toy Marks: T.P.S. (rear window deck).
Size: L: 10.5in H: 3.25in W: 4.25in (L: 27cm H: 8cm W: 11cm)
Introduced: c.1973. Est. quantity: 30,000 pieces. Scarcity rating: 3.
*Fine to Excellent (C6 to C8): $100-$135, Mint (C10): $175.*

Box text: "BATTERY OPERATED Automatic Swing and Go," "THE SWINGER." Box marks: T.P.S.

**Action:** Car goes ahead and stops with squealing brake sound. Car spins out in tight circle via underside pivoting gear and wheel mechanism then backs up before going ahead again and repeating actions. Front wheels can be positioned for direction.

# THE SWINGER (Version 2)

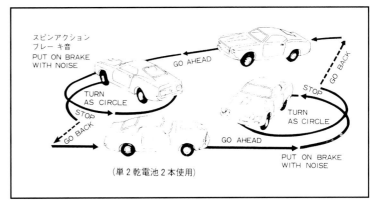

**Action:** Car goes ahead and stops with squealing brake sound. Car spins out in tight circle via underside pivoting gear and wheel mechanism then backs up before going ahead again and repeating actions. Front wheels can be positioned for direction.

Battery operated BLUE MACH 1 1970 MUSTANG MARKED "THE SWINGER," WITH RACING LOGOS AND BLUE TINTED WINDOWS. Tin with plastic wheels.
Toy Marks: T.P.S. (left rear fender).
Size: L: 10.5in H: 3.25in W: 4.25in (L: 27cm H: 8cm W: 11cm)
Introduced: c.1979. Est. quantity: 30,000 pieces. Scarcity rating: 3.
*Fine to Excellent (C6 to C8): $100-$125, Mint (C10): $150.*

# TIPPY CHOO CHOO TRAIN

TWO TRAINS ROTATING ON CYLINDRICAL CAN-LIKE BASE WITH PLASTIC DOME, CONTAINING CHILDREN'S TRAIN AND ANIMAL SCENES. Tip over windup. Tin with plastic dome and rubber smoke stacks.
Toy Marks: T.P.S. (side of toy near base).
Box text: "AUTOMATIC TIPPY CHOO CHOO TRAIN * EXCELLENT TOY FOR LITTLE CHILDREN * SAFE * EASY * DEVICE." Box marks: T.P.S.
**Action:** Tipping over toy resets weight on internal rod, which causes trains to spin and bell to ring as rod slowly drops by gravity within base of toy.
Size: H: 7in Diameter: 6in (H: 18cm Diameter: 15cm)
Introduced: c.1968. Est. quantity: 40,000 pieces. Scarcity rating: 3.
*Fine to Excellent (C6 to C8): $85-$110, Mint (C10): $145.*

## TIPPY TOY HUNTER & RABBIT

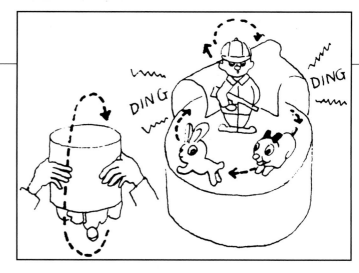

HUNTER WITH RABBIT CHASED BY DOG, ON
CYLINDRICAL CAN-LIKE BASE WITH PROPERTY
SIGNS, FOREST, AND ROCK TUNNEL. Tip over windup.
Tin. Known variations: Korean copy from Cragstan.
Toy Marks: T.P.S. (side of toy).
Box text: "tippy toy™ HUNTER & RABBIT * JUST PICK
ME UP * TURN ME OVER * WATCH ME GO GO GO!"
Box marks: T.P.S. or Cragstan.
Size: H: 7in Diameter: 6.75in (H: 18cm Diameter: 17cm)
Introduced: c.1965. Est. quantity: 15,000 pieces. Scarcity
rating: 4.
*Fine to Excellent (C6 to C8): $135-$175, Mint (C10): $225.*

**Action:** Tipping over toy resets weight on internal rod,
which causes rabbit and dog to spin and bell to ring as rod
slowly drops by gravity within base of toy.

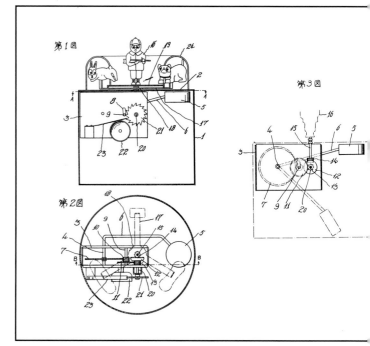

"Tippy Toy Hunter," Japanese Patent # 809173 *Courtesy of Toplay (T.P.S.) Ltd*

## TIPPY TOY TRAIN

TWO TRAIN ENGINES WITH BELL, CIRCLING ON CYLINDRICAL
CAN-LIKE BASE WITH COUNTRY SCENES, BRIDGE, AND TUN-
NEL. Tip over windup. Tin with rubber smoke stacks.
Toy Marks: T.P.S. (side of toy near base).
Size: H: 6in Diameter: 6.75in (H: 15cm Diameter: 17cm)
Introduced: c.1965. Est. quantity: 50,000 pieces. Scarcity rating: 2.5.
*Fine to Excellent (C6 to C8): $75-$95, Mint (C10): $125.*

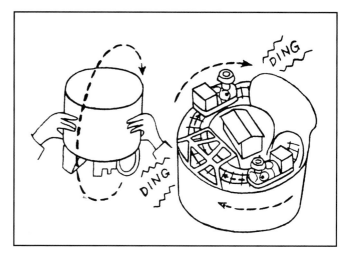

Box text: "tippy toy™ train" or "tippy train," "JUST PICK ME UP *
TURN ME OVER * WATCH ME GO GO GO!" Box marks: T.P.S.,
Cragstan, NGS.

**Action:** Tipping over toy resets weight on a rod, which causes
trains to spin and bell to ring as rod slowly drops by gravity within
base of toy.

# TODDLE'N BABY

Box text: "MECHANICAL
TODDLE'N BABY." Box
marks: T.P.S.

**Action:** Baby with shock
of hair sways back and
forth, causing baby to
toddle while its arms move
up and down.

Windup WALKING (TODDLING) BABY WITH WAVING ARMS,
with fixed key. Tin frame with vinyl body and felt dress and bonnet.
Toy Marks: None.
Size: H: 6.25in W: 4in D: 2in (H: 16cm W: 10cm D: 5cm)
Introduced: c.1965. Est. quantity: 12,000 pieces. Scarcity rating: 3.5.
*Fine to Excellent (C6 to C8): $85-$110, Mint (C10): $145.*

# TOUCHDOWN PETE

Windup FOOTBALL PLAYER WHO RUNS WITH THE BALL, with fixed key. Tin with rubber wheels.
Toy Marks: T.P.S. (inside left leg) or LINEMAR (inside left leg).
Size: L: 5.5in H: 4.75in W: 2.25in (L: 14cm H: 12cm W: 6cm)
Introduced: c.1956. Est. quantity: 50,000 pieces. Scarcity rating: 3.5.
*Fine to Excellent (C6 to C8): $250-$325, Mint (C10): $425.*

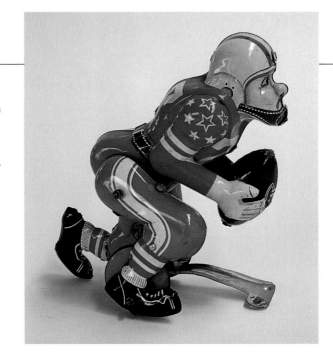

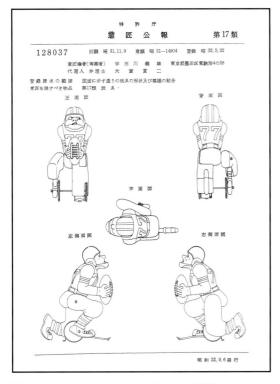

"Touchdown Pete," patent applied for 1956, granted 1957. # 128037. *Courtesy of Toplay (T.P.S.) Ltd.*

Box text: "MECHANICAL TOUCHDOWN PETE." Box marks: T.P.S. or LINEMAR. Box variation: GRAY CUP CHAMPION, add 20%. *Photo courtesy of New England Auction Gallery.*
**Action:** Football player in the kneeling position stands up at the waist as he moves forward. Right leg is attached to wheel and moves as if running. Player bends down and turns via lever extending down from base, which lifts one wheel from the surface.

# TRICYCLE TOT

Windup LITTLE GIRL PUSHING TRICYCLE, with fixed key. Tin with vinyl head and rubber heel on shoe.
Toy Marks: T.P.S. (inside right leg on pants).
Size: L: 5.5in H: 4in W: 2.75in (L: 14cm H: 10cm W: 7cm)
Introduced: c.1960. Est. quantity: 12,000 pieces. Scarcity rating: 3.5.
*Fine to Excellent (C6 to C8): $135-$175, Mint (C10): $225.*

Box text: "MECHANICAL TRICYCLE TOT." Box marks: T.P.S.

**Action:** Girl with hands on handlebars and left foot on tricycle bends forward and pushes off with right foot, causing tricycle to move forward. Jointed at shoulders, left knee, and hip.

# TROMBONE PLAYER

Windup BLOND or BLACK HAIRED TROMBONE PLAYER IN BAND UNIFORM, with fixed key. Tin.
Toy Marks: T.P.S. or LINEMAR (back below key).
Size: H: 5.25in W: 2in D: 3in (H: 13.5cm W: 5cm D: 7.5cm)
Introduced: c.1956. Est. quantity: 6000 pieces of each style. Scarcity rating: 4.
*Fine to Excellent (C6 to C8): $200-$250, Mint (C10): $300.*

Box text: "MECHANICAL TROMBONE PLAYER." Box marks: T.P.S. or LINEMAR.

**Action:** Right arm goes up and down attached to trombone which moves in and out of mouth, giving the appearance of trombone slide. Body vibrates and moves.

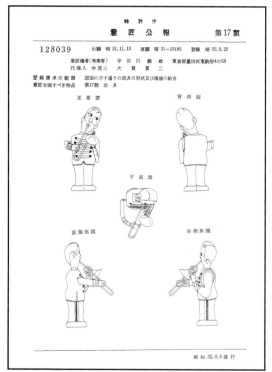

"Trombone Player," patent applied for 1956, granted 1957. # 128039. *Courtesy of Toplay (T.P.S.) Ltd.*

## TUMBLING CHIMP

Box text: "MECHANICAL TUMBLING CHIMP." Box marks: T.P.S.

**Action:** Chimp rotates at the arms and does handstand before tumbling through in a somersault fashion, landing on his feet, then repeats actions.

Windup LONG ARMED CHIMPANZEE TUMBLING WITH HANDSTANDS, with fixed key. Tin with red felt pants and hat.
Toy Marks: T.P.S. (back of right shoe).
Size: H: 4.5in W: 4in D: 2.5in (H: 11.5cm W: 10cm D: 6.5cm)
Introduced: c.1960. Est. quantity: 10,000 pieces. Scarcity rating: 4.
*Fine to Excellent (C6 to C8): $225-$300, Mint (C10): $375.*

## TWEET TWEET BIRD

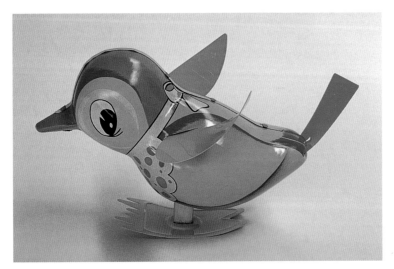

Windup YELLOW SONG BIRD WITH PLASTIC TAIL FEATHERS AND WINGS, with fixed key. Tin with plastic wings. Known variations: Also seen as "Jumping Bird" with spring jumping mechanism.
Toy Marks: T.P.S. (bottom left side of bird).
Size: L: 7.5in H: 4.75in W: 5.25in (L: 19cm H: 12cm W: 13.5cm)
Introduced: c.1965. Est. quantity: 12,000 pieces. Scarcity rating: 3.5.
*Fine to Excellent (C6 to C8): $65-$80, Mint (C10): $100.*

Box text: "MECHANICAL TWEET TWEET BIRD - SHE SINGS A SONG." Box marks: T.P.S.

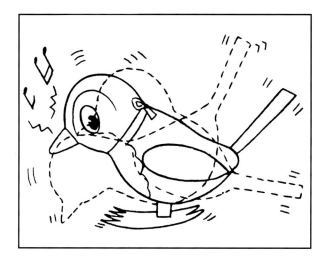

**Action:** Bird rocks back and forth in a pecking motion while emitting chirping sounds.

# TWO GUN TEX

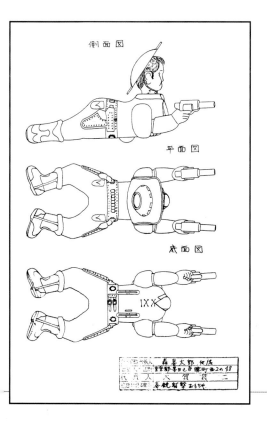

Windup COWBOY CRAWLING ON GROUND WHILE FIRING GUNS IN EACH HAND, with fixed key. Tin with vinyl head and felt cowboy hat.
Toy Marks: T.P.S. (underside of right leg near belt).
Size: L: 10.75in H: 4.75in W: 4.25in (L: 27cm H: 12cm W: 10.5cm)
Introduced: c.1960. Est. quantity: 15,000 pieces. Scarcity rating: 4.
*Fine to Excellent (C6 to C8): $275-$350, Mint (C10): $450.*

Box text: "TWO GUN TEX * ANIMATED BY WIND UP MOTOR * ROSKO 0320." Box marks: T.P.S., Rosko.

**Action:** Cowboy moves forward in crawling fashion due to off-center wheels. Descending lever causes him to turn to the left. Head turns left and right while arms raise in gun firing action with clicking sound.

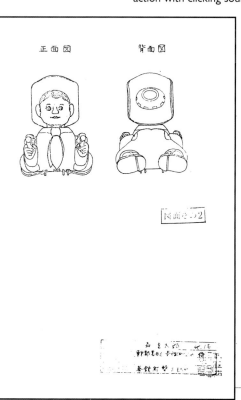

Multiple outline views of "Two Gun Tex." *Courtesy of Toplay (T.P.S.) Ltd.*

# U-TURN CATERPILLAR BULLDOZER

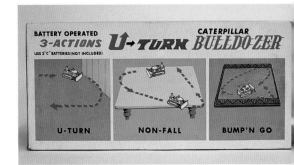

Battery operated BUMP AND GO BULLDOZER WITH U-TURN ACTION. Tin with plastic base, seat, and figure. Rubber tread.
Toy Marks: T.P.S. (front of bulldozer blade).
Size: L: 9.5in W: 5in H: 4.75in (L: 24cm W: 13cm H: 12cm)
Introduced: c.1973. Est. quantity: 40,000 pieces. Scarcity rating: 3.5.
*Fine to Excellent (C6 to C8): $125-$155, Mint (C10): $200.*

Box text: "BATTERY OPERATED U-TURN CATERPIL-LAR BULLDOZER," "3-ACTIONS * U-TURN * NON-FALL * BUMP' N GO." Box marks: T.P.S.

**Action:** Bulldozer moves forward with both bump and go as well as non fall actions. Bumping causes pedestal to lower from base, raising bulldozer up to do a U-turn and go in a different direction. Complete with engine noise and popping exhaust stack.

# U-TURN LOCO

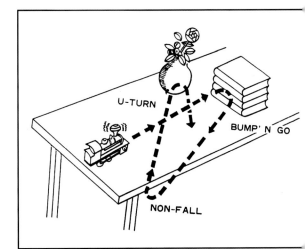

Battery operated BUMP AND GO WESTERN LOCOMOTIVE WITH U-TURN ACTION. Tin with plastic wheels. *Photo courtesy of Japan Toys Museum Foundation.*
Toy Marks: T.P.S. (rear of coal tender).
Size: L: 10.5in H: 5.5in W: 2.5in (L: 26.5cm H: 14cm W: 6.5cm)
Introduced: c.1973. Est. quantity: 15,000 pieces. Scarcity rating: 4.
*Fine to Excellent (C6 to C8): $100-$135, Mint (C10): $175.*

**Action:** Loco moves forward with both bump and go as well as non fall actions. Bumping causes pedestal to lower from base, raising train up to do a U-turn in a different direction.

# U-TURN RALLY

Battery operated RALLY TYPE PORSCHE 911-S, MARKED U-TURN, WITH PEDESTAL IN BASE. Tin with plastic steering wheel. Known variations: TV Service car. Same U-turn action with TV cameraman on roof of car. *Japan Toys Museum Foundation Collection.*
Toy Marks: T.P.S. (passenger dashboard).
Size: L: 9.75in H: 3.25in W: 4in (L: 25cm H: 8cm W: 10cm)
Introduced: c.1975. Est. quantity: 12,000 pieces. Scarcity rating: 4.
*Fine to Excellent (C6 to C8): $125-$155, Mint (C10): $200.*

Box text: "BATTERY OPERATED U-TURN RALLY," "3-ACTIONS * U-TURN * NON-FALL * BUMP' N GO." Box marks: T.P.S.

**Action:** Porsche moves forward with both bump and go as well as non fall actions. Pedestal lowers from base, raising car up. Car does U-turn then pedestal raises (car lowers) and car goes forward again.

# UNCLE GEBA-GEBA

Windup WALKING UNCLE GEBA-GEBA, JAPAN'S NTV CHARACTER, with fixed key. Plastic with tin overlay. *Tokyo Robot Collection.*
Box text: "GEBA-GEBA OJISAN (in Japanese) NTV Studio Lotus." Box marks: T.P.S.
**Action:** Uncle Geba-Geba walks slowly forward with fixed arms and hat.
Size: H: 5.5in W: 2.75in D: 2.25in (H: 14cm W: 7cm D: 5.5cm)
Introduced: c.1972. Est. quantity: 6000 pieces. Scarcity rating: 5.
*Fine to Excellent (C6 to C8): $300-$375, Mint (C10): $500.*

# URAN BOUNCING BALL

Windup TETSUWAN ATOM'S (ASTROBOY'S) SISTER, URAN, BOUNCING BALL WITH HER HAND, with fixed key. Tin with vinyl head.
**Action:** Uran bounces ball between floor and her right hand. Both arms move as ball moves up and down, attached to body mechanism by wire.
Size: H: 5.5in W: 2.75in D: 4.75in (H: 14cm W: 7cm D: 12cm)
Introduced: c.1965. Est. quantity: 30,000 pieces. Scarcity rating: 5.
*Fine to Excellent (C6 to C8): $1,500-$2,000, Mint (C10): $2,600.*

Box text: "URAN CHILD BOUNCING BALL" (Japanese). Box marks: TADA, Atom.

Toy Marks: TADA -Atom (left rear pants under skirt), T.P.S. (top of ball). The T.P.S. mark may be painted over.

## VIOLINIST

Windup VIOLIN PLAYER WITH MOVING ARM AND HEAD, with fixed key. Tin.
Toy Marks: LINEMAR (on back below key).
Size: H: 5.25in W: 2.25in D: 2.75in (H: 13.5cm W: 5.5cm D: 7cm)
Introduced: c.1956. Est. quantity: 6,000 pieces. Scarcity rating: 4.
*Fine to Excellent (C6 to C8): $200-$250, Mint (C10): $300.*

Box text: "MECHANICAL VIOLINIST J-1408." Box marks: LINEMAR.

**Action:** Man holds violin in left hand while right arm moves bow back and forth across the violin. Head moves left and right while body moves from vibrating action.

# VOLKSWAGEN CAR ON HIGHWAY

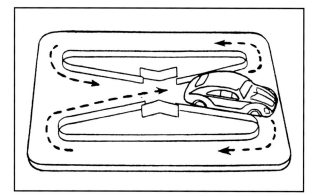

Windup 7cm (2.75in) VOLKSWAGEN ON PLATFORM BASE WITH HIGHWAY SCENES AND VOLKSWAGENS. Separate key. Tin.
Toy Marks: T.P.S. (base, lower left side), Hirata (base, bottom right side).
Size: 9.25in x 5.5in platform (23.5cm x 14cm platform)
Introduced: c.1967. Est. quantity: 20,000 pieces. Scarcity rating: 4.
*Fine to Excellent (C6 to C8): $175-$200, Mint (C10): $250.*

Box text: "MECHANICAL VOLKSWAGEN CAR ON HIGH WAY." Box marks: T.P.S.

**Action:** Volkswagen goes around highway platform hooked on guide rail in figure eight route.

# WAGON FANTASY LAND a.k.a. FAIRY TALE LAND TAXI

Windup BEETLE TYPE BUG PULLING TWO SQUIRRELS ON WAGON WITH MONKEY DRIVER, with fixed key. Tin with rubber wheels.
Toy Marks: T.P.S. (leaf wagon, rear).
Size: L: 11.75in H: 2.75in W: 4.25in (L: 30cm H: 7cm W: 11cm)
Introduced: c.1960. Est. quantity: 30,000 pieces. Scarcity rating: 3.
*Fine to Excellent (C6 to C8): $200-$250, Mint (C10): $350.*

Box text: "MECHANICAL TOY WAGON 'FANTASY LAND' * LOTS OF ACTION - LOTS OF FUN * ROSKO 1510." Box marks: T.P.S., Rosko.

Known box variation: "FAIRY TALE LAND TAXI, AN ANIMATED WIND-UP TOY", marked Cragstan, add 15%.

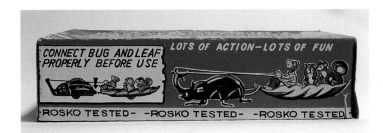

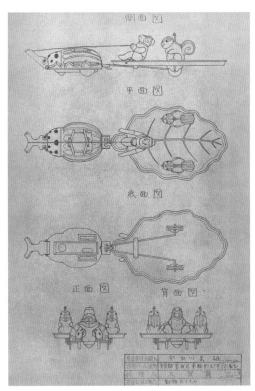

**Action:** Squirrels rotate and monkey rocks back and forth while holding on to reins attached to bug's nose. Bug's head moves up and down causing monkey to rock.

*Courtesy of Toplay (T.P.S.) Ltd.*

# WALKING CHOO CHOO

Windup CHOO CHOO WITH WALKING WHEELS, with fixed key. Tin with plastic wheels.
Toy Marks: T.P.S. (rear of engine).
Size: L: 4.75in W: 3.25in H: 3.5in (L: 12cm W: 8cm H: 9cm)
Introduced: c.1967. Est. quantity: 20,000 pieces. Scarcity rating: 3.5.
*Fine to Excellent (C6 to C8): $50-$70, Mint (C10): $100.*

Box text: "MECHANICAL WALKING CHOO CHOO." Box marks: T.P.S.

**Action:** Front wheels and hat-shaped smoke stack pivot back and forth. Rear wheels pivot and lift alternately, causing locomotive to walk forward.

# WALKING FROG - 'KEROYON'

Box text: "MECHANICAL WALKING FROG." Box marks: T.P.S.

**Action:** Frog sways back and forth causing its stationary base to move in walking manner, while its arms move up and down.

Windup WALKING JAPANESE CHARACTER FROG - "KEROYON," WITH WAVING ARMS, with fixed key. Tin with vinyl face and arms.
Toy Marks: T.P.S. (below key).
Size: H: 7in W: 3.5in D: 2.75in (H: 18cm W: 9cm D: 7cm)
Introduced: c.1967. Est. quantity: 6,000 pieces. Scarcity rating: 5.
*Fine to Excellent (C6 to C8): $550-$700, Mint (C10): $900.*

# WALKING PENGUIN

Windup WALKING PENGUIN WITH QUACKING SOUND AND UPLIFTED FLAPPING WINGS, with fixed key. Tin. Known variations: "WALKING DUCK" is the same toy with duck lithography.
Toy Marks: Made in Japan (front left side near base).
Size: H: 4.75in W: 3.25in D: 4.25in (H: 12cm W: 8cm D: 11cm)
Introduced: c.1959. Est. quantity: 6000 pieces. Scarcity rating: 5.
Fine to Excellent (C6 to C8): $275 $325, Mint (C10): $375.

Box text: "MECHANICAL ACTION WALKING PENGUIN * Quacking and Flapping Wings." Box marks: Made in Japan.

**Action:** Penguin moves forward and turns in a walking or waddling fashion via irregular shaped wheels. Penguin stops and begins to make quacking sound while flapping its wings and opening and closing its lower beak, before repeating actions.

# WALKING ROBOT

Windup CHROME PLATED WALKING AND SPARKING ROBOT, with fixed key. Chrome plated plastic.
Toy Marks: Made in Japan (molded on back of head).
Size: H: 5in W: 2.75in D: 2.25in (H: 13cm W: 7cm D: 5cm)
Introduced: c.1971. Est. quantity: 30,000 pieces. Scarcity rating: 4.
*Fine to Excellent (C6 to C8): $60-$75, Mint (C10): $100.*

Box text: "MECHANICAL WALKING ROBOT * SPARKING * NO.2022." (In English and Japanese). Box marks: T.P.S.

**Action:** Robot walks slowly while shooting sparks from internal sparking mechanism in chest.

# WALKING TURTLE

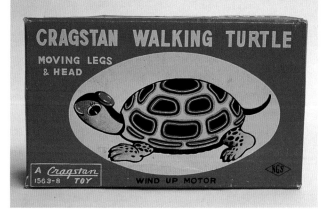

Windup WALKING TURTLE WITH MOVING LEGS AND ARMS, with fixed key. Tin with rubber tail.
Toy Marks: T.P.S. (left rear shell).
Size: L: 7in W: 4in H: 2.5in (L: 18cm W: 10cm H: 6.5cm)
Introduced: c.1965. Est. quantity: 20,000 pieces. Scarcity rating: 3.5.
*Fine to Excellent (C6 to C8): $100-$125, Mint (C10): $165.*

Box text: "CRAGSTAN WALKING TURTLE * MOVING LEGS & HEAD * WIND UP MOTOR * Cragstan 1563-8." Box marks: Cragstan, NGS.

**Action:** Turtle walks very slowly as legs move alternately in realistic motion. Head turns to left and right.

# WARBLER IN CAGE

Windup SINGING TIN BIRD IN CAGE, with fixed key.
Tin with plastic cage.
Size: H: 6.25in (H: 16cm)
Introduced: c.1960. Est. quantity: 6,000 pieces.
Scarcity rating: 5.
*Fine to Excellent (C6 to C8): $175-$225, Mint (C10): $300.*
Box text: "MECHANICAL WARBLER IN CAGE." Box
marks: T.P.S., Frankonia.

**Action:** Bird attached to mechanism by rod hops from feeding dish on
bottom of cage to perch. Voice box in mechanism creates warbling effect for
bird.

# WESTERN TRAIN

Windup ARTICULATED 3 CAR TRAIN WITH WESTERN COWBOY AND INDIAN
COMIC FACES AT THE WINDOWS OF THE ENGINE AND EACH CAR. Fixed key.
Tin with rubber stack and wheels.
Toy Marks: T.P.S. (rear of last car).
Size: L: 11.5in H: 2in W: 1.25in (L: 29cm H: 5cm W: 3cm)
Introduced: c.1967. Est. quantity: 30,000 pieces. Scarcity rating: 3.5.
*Fine to Excellent (C6 to C8): $75-$100, Mint (C10): $125.*

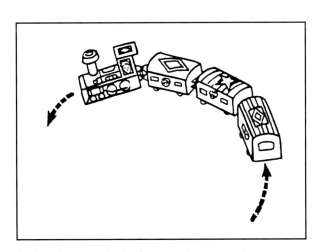

**Action:** Train goes in circular route due to oversize
right wheel on the engine. Jointed cars follow in zigzag
fashion.

# WESTERN TRAIN (Battery Operated)

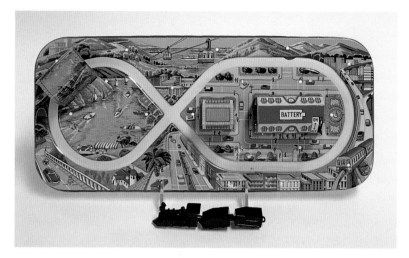

Battery operated WESTERN STYLE PLASTIC STEAM LOCOMOTIVE TRAVELS AROUND FIGURE EIGHT TRACK ON PLATFORM BASE WITH CITY AND COUNTRY SCENES, TUNNEL, AND BUILDINGS. Tin with plastic train.
Toy Marks: T.P.S. (lower left corner of base), Hirata (lower left corner of base).
Size: L: 15.25in W: 6.75in H: 2in (L: 38.5cm W: 17cm H: 5cm)
Introduced: c.1968. Est. quantity: 20,000 pieces. Scarcity rating: 4.
*Fine to Excellent (C6 to C8): $150-$200, Mint (C10): $250.*

Box text: "BATTERY OPERATED WESTERN TRAIN * VIBRATING ACTION."
Box marks: T.P.S.

**Action:** Vibrating action causes 13cm (5in) train to move around track. Loco and cars have brush type nap underneath to keep them moving forward only. Battery housed in building with on-off switch.

Detail of underside of train showing brush type nap.

# WHISTLING TOYLAND CHOO CHOO

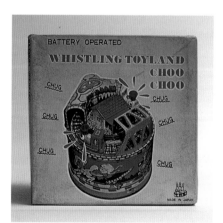

BATTERY POWERED TRAINS CIRCLING ON CYLINDRICAL CAN-LIKE BASE WITH CHILDREN'S TRAIN AND ANIMAL SCENES, ALONG WITH BRIDGE, STATION, AND TUNNEL. Tin with red plastic smokestacks.
Toy Marks: T.P.S. (on side near base).
Size: H: 6in Diameter: 6in (H: 15cm Diameter: 15cm)
Introduced: c.1968. Est. quantity: 6,000 pieces. Scarcity rating: 4.5.
*Fine to Excellent (C6 to C8): $450-$575, Mint (C10): $750.*

Box text: "BATTERY OPERATED WHISTLING TOYLAND CHOO CHOO." Box marks: T.P.S.
**Action:** Both engines travel around through bridge and tunnel with lights flashing in their smoke stacks. Engine sounds come from within base.

# WILD WHEELING DUNE BUGGY

Battery operated ORANGE DUNE BUGGY, WITH ADJUSTABLE FRONT WHEELS, THAT DOES WHEELIES. Tin with plastic fenders and driver, rubber tires.
Toy Marks: T.P.S. (passenger seat).
Size: L: 11 W: 5.5in H: 4.75in (L: 28cm W: 14cm H: 12cm)
Introduced: c.1970. Est. quantity: 20,000 pieces. Scarcity rating: 3.
*Fine to Excellent (C6 to C8): $75-$100, Mint (C10): $125.*

Box text: "BATTERY OPERATED WILD WHEELING DUNE BUGGY," "WHEEL STAND ACTION!! * IT DOES A REAL WHEEL STAND! * TURNS AROUND, COME DOWN! * STOP AND GO ACTION! * ROARING ENGINE SOUND! * STEER IT YOURSELF!" Box marks: T.P.S., MEGO.

**Action:** Buggy goes forward. Lever with fifth wheel drops down from base causing front wheels to raise and spin buggy around. Buggy lowers and goes forward again before repeating actions. Front wheels can be positioned for direction.

# WORLD CHAMPION AUTO RACE

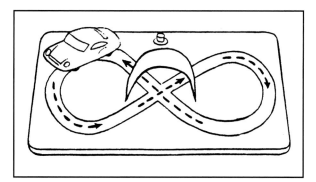

Windup 6cm (2.25in) RACE CAR #77, ON PLATFORM BASE WITH FIGURE EIGHT RACE TRACK, RACE TRACK SCENES, AND ARCH, with base mounted key. Tin.
Toy Marks: T.P.S. (bottom right side of base), Hirata (lower right side of base).
Size: 9.25in x 5.5in platform (23.5cm x 14cm platform)
Introduced: c.1965. Est. quantity: 30,000 pieces. Scarcity rating: 3.5.
*Fine to Excellent (C6 to C8): $125-$150, Mint (C10): $200.*

Box text: "MECHANICAL WORLD CHAMPION AUTO RACE." Box marks: T.P.S.

**Action:** Race car goes around racetrack platform hooked on guide rail, in figure eight pattern.

## ZIG-ZAG SPEED RACE

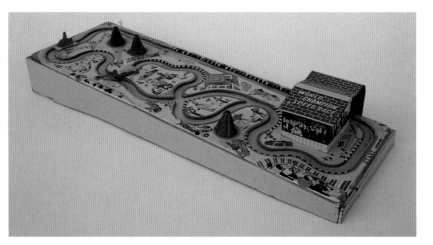

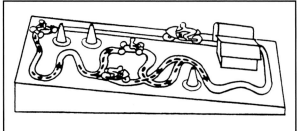

TWO GO-CARTS AND TWO MOTORCYCLES RACE AROUND WINDING TRACK ON SLOPED PLATFORM BASE WITH RACE TRACK SCENES AND TUNNEL BUILDING MARKED "WORLD CHAMPION SPEED RACE." Battery operated. Tin with plastic go-carts and motorcycles.
Toy Marks: T.P.S. (corner near building).
Size: L: 20in W: 6.75in H: 3.25in (L: 51cm W: 17cm H: 8.5cm)
Introduced: c.1967. Est. quantity: 6000 pieces. Scarcity rating: 5.
*Fine to Excellent (C6 to C8): $185-$240, Mint (C10): $300.*

Box text: "BATTERY OPERATED ZIG-ZAG SPEED RACE W/ RACE CART AND MOTOR CYCLE * VIBRATING ACTION."
Box marks: T.P.S.

**Action:** Vibrating action causes 3.5cm (1.5in) go-carts and motorcycles to move around sloped track. Alternate path may be selected by sliding lever on base. Racers have brush type nap underneath to keep them moving forward only. Batteries housed in platform base with on-off switch.

# MISSING TOYS

These two drawings from Toplay (T.P.S.) Ltd. represent toys that I have not been able to find. They were designed around 1959-1960. Maybe you have seen them!

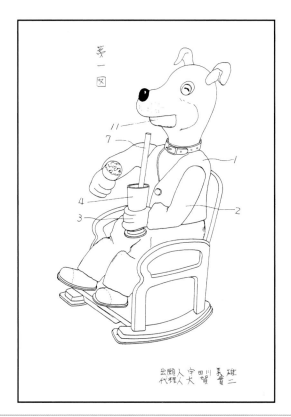

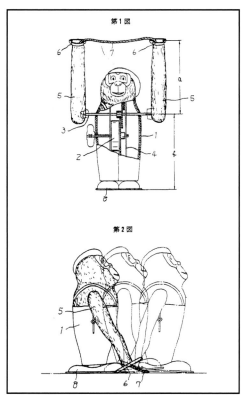

# TOY RESOURCES

These businesses, who contributed photographs for this publication, sell and/or repair antique and collectible toys. T.P.S. toys are usually found on their sale lists or in their regular auctions.

Randy Ibey of Randy's Toy Shop, specializing in the repair and restoration of antique toys and boxes. Randy's Toy Shop, 165 North 9th Street, Noblesville, Indiana.

Debby and Marty Krim's New England Auction Gallery. Their regular auction normally contains a fine selection of T.P.S. toys. New England Auction Gallery, PO Box 2273, West Peabody, Massachusetts 01969.

Barb and Herb Smith of Smith House Toys. Their regular auction also normally contains a fine selection of T.P.S. toys. Smith House Toys, PO Box 336, Eliot, Maine 03903.

# BIBLIOGRAPHY

Japan Toy International Trade Fair Association. *1st Japan International Toy Fair Catalog*. 1962.

Japan Toys Museum Foundation, Shigeru Mozuka, Manager; Mitsuo Tsukuda, Owner and Founder. Tokyo, Japan.

Kumagi, Nobuo. *'50s Japanese Mechanical Tin Toys*. 1980.

O'Brien, Richard. "T.P.S." Donald Hultzman, contributor. *Collecting Toys*. Books Americana, 1997.

Smith, Herb. *Smith House Toys 1998 Price Guide*. Herb Smith, 1998.

Tada, Toshikatsu. *The Toy Museum Vol. 1, 2, and 3*. Kyoto: Kyoto Shoin, 1992.

Takayama, Toyoji. *Nostalgic Tin Toys Vol. 1, 2 and 3*. Kyoto: Kyoto Shoin, 1989.

The Tin Toy Museum and collection of Toyoji Takayama,

# INDEX OF TOY NAMES BY CATEGORY

All toys included in this work are listed in alphabetical order by the actual "Toy Name." To assist in identifying these actual names, this index contains a listing of toys by subject category with thumbnail photographs. Toys are listed by each of the following categories:

Airplanes and Helicopters
Amusement Park Toys
Animal Toys
Character Toys
Circus and Clown Toys
Climbing Toys
Cyclist Toys

Musician Toys
Platform Base Toys
Skating Toys
Space Toys
Sports Toys
Train Toys
Vehicle Toys
Miscellaneous Toys

Underneath each photo is the correct name of the toy. The detailed information regarding each toy is included in the alphabetical listing.

## AIRPLANES AND HELICOPTERS

DREAMLAND AIR PORT

FLYING SKY PATROL

GRUMMAN F-14A JET FIGHTER TOMCAT

HAPPY PLANE

HELICOPTER ON AIRFIELD

HELICOPTER WITH AUTOMOBILE

JOLLY PLANE

MERRY PLANE WITH BELL

MITSUBISHI ZERO FIGHTER

POLICE HELICOPTER

SKY CIRCUS SPAD XIII

STUNT PLANE

SUPER FLYING HELICOPTER

SUPER FLYING POLICE
HELICOPTER

SUPER FLYING TRAFFIC
CONTROL HELICOPTER

SUPER THUNDER WOLF

TAKE OFF AIRPORT

# AMUSEMENT PARK TOYS

CONNY ISLAND SCOOTER

DREAM LAND BUS in
MAGIC TUNNEL

DREAMLAND MIDGET
CHOO CHOO

DREAMLAND WITH BELL

FERRIS WHEEL TRUCK

LOOP THE LOOP COASTER

MERRY-GO-ROUND TRUCK

MERRY PLANE WITH BELL

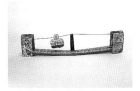

PLAYLAND CABLE CAR
version 1

PLAYLAND CABLE CAR
version 2

PLAYLAND SCOOTER

# ANIMAL TOYS

ANIMAL BARBER SHOP

ANIMAL'S PLAYLAND

BALL PLAYING GIRAFFE

BEAR GOLFER

BEAR PLAYING BALL

BUNNY FAMILY PARADE

BUSY BUG

CANDY LOVING CANINE

CAT & MOUSE IN SHOE
WITH VOICE

CHAMP ON ICE

CIRCUS ACROBATIC SEAL
AND BALL

CIRCUS ELEPHANT

CIRCUS MONKEY

CIRCUS SEAL

COCK-A-DOODLE

COW HOUSE WITH VOICE

DOG HOUSE WITH VOICE

DUCK AMPHIBIOUS TAXI

DUCK FAMILY PARADE

DUCK THE MAIL MAN

EDUCATIONAL PET POOCH

FAMILY GIRAFFE LOCO

FISHING BEAR

FISHING MONKEY ON
WHALES

FLYING BIRDS WITH VOICE

HAPPY HIPPO

HAPPY SKATERS - BEAR

HAPPY SKATERS - RABBIT

HUNGRY CAT

HUNGRY WHALE

JOLLY WIGGLING SNAKE

JUGGLING DUCK & HIS
FRIENDS

JUMPING CAT

JUMPING SQUIRREL

LADY-BUG FAMILY PARADE
(a.k.a. LADY-BUGS)

LADY BUG & TORTOISE
WITH BABIES

LUCKY MONKEY PLAYING
BILLIARDS

MAMA KANGAROO
WITH PLAYFUL BABY
IN HER POUCH

MERRY PENGUIN

MIDGET LADY-BUG

MIGHTY MIDGET 'ON-TABLE' MECHANICALS

MONKEY BASKET BALL
PLAYER

MONKEY GOLFER

MOUSE CHASER

MOUSE FAMILY

MR. CATERPILLAR
(a.k.a. HAIRY CATERPILLAR)

OSCAR THE PERFORMING
SEAL - BALL

OSCAR THE SEAL

PAT THE PUP

PERFORMING SEAL AND
MONKEY WITH FISH

PLAYFUL CIRCUS SEALS

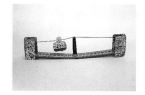

PLAYLAND CABLE CAR
version 1

PLAYLAND CABLE CAR
version 2

PUSSY CAT CHASING BUT-
TERFLY (a.k.a. CASSIUS-THE
CAT CHASING BUTTERFLIES)

RABBIT AND BEAR
PLAYING BALL

SHUTTLE ZOO TRAIN

SKIP ROPE ANIMALS

SLIM THE SEAL & his
FRIENDS

SOMERSAULT FROG

STANDING
TURTLE

SUSIE THE OSTRICH

TUMBLING
CHIMP

TWEET TWEET BIRD

WAGON FANTASY LAND
(a.k.a. FAIRY TALE LAND TAXI)

WALKING FROG - KEROYON

WALKING PENGUIN

WALKING TURTLE

WARBLER IN CAGE

# CHARACTER TOYS

CALYPSO JOE THE DRUMMER

COMICAL CLARA

GOOFY CYCLIST

GOOFY THE UNICYCLIST

GRENDIZER STAND UP
CYCLE

JUGGLING POPEYE AND
OLIVEOYL

MANHATTAN BANK
(a.k.a. KING KOIN BANK)

MICKEY MOUSE CYCLIST

MICKEY MOUSE
ROLLER SKATER

MICKEY MOUSE THE
UNICYCLIST

MOUNTED CAVALRYMAN
WITH CANNON

PANGO-PANGO
AFRICAN DANCER

PLUTO THE UNICYCLIST

POP EYE PETE

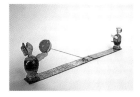

POPEYE AND OLIVE OYL
PLAYING BALL

POPEYE CYCLIST

POPEYE SKATER

POPEYE THE BASKET-
BALL PLAYER

POPEYE UNICYCLIST

SHY ANNE INDIAN SKATER

TETSUJIN 28 GO (GIGANTOR)

TETSUWAN MIGHTY ATOM
(ASTROBOY)

TETSUWAN ATOM
(ASTROBOY) FLYING
SAUCER FAMILY

TWO GUN TEX

UNCLE GEBA-GEBA

URAN BOUNCING BALL

WALKING FROG - KEROYON

# CIRCUS AND CLOWN TOYS

BOBO THE JUGGLING
CLOWN

CIRCUS ACROBATIC SEAL
AND BALL

CIRCUS BUGLER

CIRCUS CLOWN

CIRCUS CYCLIST

CIRCUS ELEPHANT

CIRCUS MONKEY

CIRCUS PARADE (a.k.a.
ELEPHANT CIRCUS PARADE)

CIRCUS SEAL

CLIMBO THE CLIMBING
CLOWN

CLOWN JALOPY CYCLE

CLOWN JUGGLER WITH BALL

CLOWN JUGGLER
WITH MONKEY

CLOWN MAKING LION JUMP
THRU FLAMING HOOP

CLOWN ON ROLLER SKATE

CLOWN ON ROLLER
SKATE-WHITE FACE

CLOWN TRAINER AND HIS
ACROBATIC DOG (a.k.a. CLEO
CLOWN THE DOGS)

CLOWN'S POPCORN TRUCK

DANCING CLOWN

DUCK AMPHIBIOUS TAXI

EDUCATIONAL PET POOCH

HAPPY, THE VIOLINIST
(Version 1)

HAPPY, THE VIOLINIST
(Version 2)

JOE THE ACROBAT

JOE THE XYLOPHONE
PLAYER

JUGGLING CLOWN
(WITH APPLES)

JUGGLING CLOWN
(WITH BALL)

JUGGLING DUCK &
HIS FRIENDS

MAGIC CIRCUS

OSCAR THE PERFORMING
SEAL - BALL

OSCAR THE SEAL

PERFORMING SEAL AND
MONKEY WITH FISH

PLAYFUL CIRCUS SEALS

ROLLER SKATING CIRCUS
CLOWN

SAMSON THE STRONG MAN

SKATING CHEF - BLACK

SKIPPY, THE TRICKY CYCLIST

SLIM THE SEAL & his FRIENDS

SUSIE THE OSTRICH

TUMBLING CHIMP

# CLIMBING TOYS

CLIMBING LINESMAN

CLIMBING PIRATE

CLIMBO THE CLIMB-
ING CLOWN

FIREMAN V

LOOP-THE-LOOP COASTER

MANHATTAN BANK
(a.k.a. KING KOIN BANK)

MOUNTAIN CLIMBER

PLAYFUL CIRCUS SEALS

ROBOT MACHINE

SPACE STATION CHASE

THE MONKEY TREE

# CYCLIST TOYS

CIRCUS CLOWN

CIRCUS CYCLIST

GAY 90'S CYCLIST

GOOFY CYCLIST

GOOFY THE UNICYCLIST

MICKEY MOUSE CYCLIST

MICKEY MOUSE THE
UNICYCLIST

PLUTO THE UNICYCLIST

POPEYE CYCLIST

POPEYE UNICYCLIST

SKIPPY, THE TRICKY CYCLIST

## MUSICIAN TOYS

BANJOIST

CALYPSO JOE THE DRUMMER

CIRCUS BUGLER

HAPPY, THE VIOLINIST
(Version 1)

HAPPY, THE VIOLINIST
(Version 2)

JOE THE XYLOPHONE
PLAYER

TROMBONE PLAYER

VIOLINIST

## PLATFORM BASE TOYS

BUSY CHOO CHOO

BUSY F.D. LADDER TRUCK

BUSY MOUSE

CATERPILLAR BULLDOZER

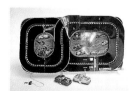

CHAMPION AUTO RACE
TOURNAMENT

COMBAT TANK ON
BATTLE FRONT

CONNY ISLAND SCOOTER

DREAM LAND BUS in
MAGIC TUNNEL

DREAM SUPER EXPRESS-
THE HIKARI

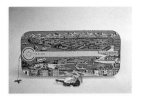

DREAMLAND AIR PORT

FIGURE 8 HIGHWAY

FIRE DEPT LADDER TRUCK

HELICOPTER ON AIRFIELD

HIGHWAY PATROL CAR

HIGHWAY SET

HIKARI EXPRESS RAIL BOARD

LOCO CAB OVERLAND
TRAFFIC GAME

MAGIC CHOO CHOO

MAGIC CROSS ROAD

MAGIC GREYHOUND BUS

MAGIC TUNNEL

MOUSE RACE CAT

PLAYFUL PUPPY

PLAYLAND SCOOTER

POLICE CAR CHASE

RAILROAD SET

ROAD RACE

SHUTTLE ZOO TRAIN

SHUTTLING LOCOMOTIVE

SIGHTSEEING BUS

SPORTS CAR RACE

TAKE OFF AIRPORT

TANK ON A BATTLEFIELD

TETSUJIN 28 GO (GIGANTOR)

TETSUWAN MIGHTY ATOM
(ASTROBOY)

VOLKSWAGEN CAR
ON HIGHWAY

WORLD CHAMPION
AUTO RACE

ZIG-ZAG SPEED RACE

# SKATING TOYS

CHAMP ON ICE

CLOWN ON ROLLER SKATE

CLOWN ON ROLLER
SKATE-WHITE FACE

HAPPY SKATERS - BEAR

HAPPY SKATERS - RABBIT

MICKEY MOUSE ROLLER
SKATER

POPEYE SKATER

ROLLER SKATING
CIRCUS CLOWN

SHY ANNE INDIAN
SKATER

SKATING CHEF

SKATING CHEF - BLACK

# SPACE AND ROBOT TOYS

APOLLO SPACE PATROL
WITH SATELLITE SHIP

BRAVE HOVERCRAFT

FLASH SPACE PATROL

FLYING AIR CAR
(a.k.a. HOVER CRAFT)

LUNA HOVERCRAFT

MERCURY EXPLORER

MINI WALKING ROBOT

MISSILE ROBOT

MOON EXPRESS

MOON PATROL

MR. SIGNAL

RADAR MISSILE ROBOT

ROBOT MACHINE

SATELLITE FLEET

SPACE EXPLORER WITH
FLYING SPACE MAN

SPACE EXPRESS

SPACE STATION CHASE

SPARKING ROBOT

WALKING ROBOT

# SPORTS TOYS

BARREL ROLL RACE CAR
SET WITH SEESAW

BEAR GOLFER

BIG LEAGUE HOCKEY
PLAYER

BOUNCING BALL DOLLY

FISHING BEAR

GIRL SKIPPING ROPE

LUCKY MONKEY PLAYING
BILLIARDS

MONKEY BASKET BALL
PLAYER

MONKEY GOLFER

SKATE BOARD

SKIP ROPE ANIMALS

SPORTS CAR RACE

SPORTS CYCLE

SUZY BOUNCING BALL
(a.k.a. PRETTY LITTLE GIRL
WITH PONYTAIL - BOUNC-
ING BALL)

TOUCHDOWN PETE

TRICYCLE TOT

WORLD CHAMPION
AUTO RACE

# TRAIN TOYS

BUSY CHOO CHOO

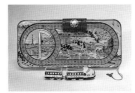

DREAM SUPER EXPRESS-
THE HIKARI

DREAMLAND MIDGET
CHOO CHOO

DREAMLAND WITH BELL

FAMILY GIRAFFE LOCO

HIKARI EXPRESS RAIL BOARD

JOLLY LOCO

LOCO CAB OVERLAND
TRAFFIC GAME

MAGIC CHOO CHOO

MAGIC CROSS ROAD

PULL TOY LOCOMOTIVES

RAILROAD SET

SHUTTLING LOCOMOTIVE

SIX TRACK TRAIN SET

TIPPY CHOO CHOO TRAIN

TIPPY TOY TRAIN

U-TURN LOCO

WALKING CHOO CHOO

WESTERN TRAIN

WESTERN TRAIN
(Battery Operated)

WHISTLING TOYLAND
CHOO CHOO

# VEHICLE TOYS

ACROBAT TEAM PORSCHE

BARREL ROLL RACE CAR SET WITH SEESAW

BIG STUNT CAR

BRAVE HOVERCRAFT

BUSY EMERGENCY CAR SERIES - FIRE

BUSY EMERGENCY CAR SERIES - POLICE

CHAMPION STUNT CAR

COLORFUL BALL BLOWING TRUCKS

CRANE TRUCK

Dax HONDA MOTOR CYCLE

DRIVE TESTER

DUNE BUGGY W/ SURF BOARD

FERRIS WHEEL TRUCK

FLYING AIR CAR

FORD MUSTANG

FORD MUSTANG STUNT CAR

GASOLINE CAR - MOBILGAS

HIGHTECHNICAL BIG RIDER

HIGHTECHNICAL RALLY

HIGHTECHNICAL RIDER

KINDERGARTEN BUS

MIGHTY MIDGET - ANTIQUE MINIATURE AUTO

MIGHTY MIDGET - ASSORTED VEHICLES

MINIATURE DUNE BUGGY

MISSILE TANK

NEWS SERVICE CAR -
WORLD NEWS

OPEN-SHUT BONNET
STUNT CAR

PULL- BACK PORSCHE TURBO

SAND CONVEYOR TRUCK -
Version 1

SAND CONVEYOR TRUCK -
Version 2

SIREN EMERGENCY SERIES

SPARKING CATERPILLAR
BULLDOZER

SPINNING PORSCHE

STOP AND GO TRUCK SERIES

THE HUSTLER

THE HUSTLER MUSTANG
HANDLING TEST CAR

THE SWINGER

THE SWINGER (version 2)

U-TURN CATERPILLAR
BULLDOZER

U-TURN RALLY

WILD WHEELING DUNE
BUGGY

# MISCELLANEOUS TOYS

DANCING COUPLE

GIRL WITH CHICKENS (a.k.a.
GIRL FEEDING CHICKENS)

GOOD COMPANION

MUSICAL HAND SEWING
MACHINE

SNAPPING SHOE

TIPPY TOY
HUNTER & RABBIT

TODDLE'N BABY